香 港 1 0 0 種
景觀樹木圖鑑

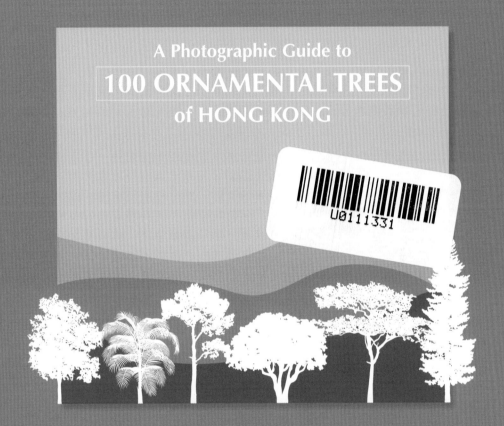

A Photographic Guide to
100 ORNAMENTAL TREES
of HONG KONG

張浩 主編

萬里機構

推薦序一

　　香港位於中國東南端，擁有超過 700 萬人口，是世上人口最稠密的地區之一。儘管香港陸地總面積只有 1114 平方公里，卻有約四成土地為郊野公園。在香港市區各處，種植了很多不同的綠化樹木與其他植物，隨着四季變化而呈現不同的景觀。香港本地植物自然資源豐富，隨着多年來城市的發展，也從外地引入了不同的植物品種，孕育了別具特色的香港園藝園境。

　　市民無論在遊覽郊野公園、水塘及自然保護區，還是穿梭於市區內的公園、海濱長廊、美化市容地帶、小型綠化空間，又或是一瞥路旁及斜坡植被，不難欣賞到觀賞價值很高的景觀樹木。這些樹木具有美觀的葉、枝、花、果，再加上多樣的季相變化，極大地豐富了香港的城市面貌，使繁忙的東方之珠散發出獨特的綠色魅力。

　　樹木作為城市美化的骨幹，優質的城市林木管理就十分重要。香港高等教育科技學院（THEi）的園藝樹藝及園境管理（榮譽）理學士課程，就是致力於培訓有關的人才。課程旨在培養學生掌握園藝及樹藝管理、園境建設及維護的理論、技術和專業的技能和知識，當中涵蓋植物科學、樹藝、環境、可持續園境管理等相關學科範疇，幫助學生履行園藝及樹藝監督、樹木評估、園境管理，以及園境發展和環境保育等領域的專業角色和任務。課程主任張浩博士，亦即本書的主編，主要研究香港城市林業及城市園藝，在植物研究和樹木評估等方面累積了多年實踐經驗。他與團隊通過進行實地調查，拍攝百種本地景觀樹木的全貌、樹幹、樹葉、樹皮、花果等照片，並搜集、考證多項文字記錄和資料，完成了此樹木圖鑑。

　　本書中除了圖文並茂地描述樹木的辨認特徵，讓讀者更加容易辨認各種景觀樹木，也提供了關於這些樹木的生動有趣之見聞，使讀者可以從輕鬆的角度認識它們。此圖鑑資料扎實豐富，配以樹木各部分的精美圖片，實為一難得的香港景觀樹木圖鑑，讓園藝師、樹藝師、景觀設計師及植物愛好者以便捷方法辨識樹木，也為景觀樹木利用和綠化地區管理和發展提供科學參考。

　　本人祝賀本圖鑑成功出版，並感謝香港特別行政區政府發展局綠化、園境及樹木管理組的支持，以及出版社編輯團隊的努力。

劉建德 教授

香港高等教育科技學院（THEi）校長

2023 年 6 月

推薦序二

　　城市樹木體現了一個地方的規劃、設計、文化和智慧。過往香港政府植樹主要是為了郊野造林、鞏固斜坡和美化市區，近年則強調樹木於都市景觀生態及保育的角色；城市森林、生態廊道、公園、行道樹、立體綠化等提供了吸碳減碳和其他生態功能的潛力，提升城市生態韌性和應對氣候變化危機的抗性。樹種多了，塌樹自然變得頻繁，大家更關注樹木栽培和維護，也孕育出更多的愛樹者和護樹者。香港曾出版《香港野外樹木圖鑑》、《香港市區常見樹木圖鑑》、《香港觀賞樹木彙編》等工具書，也有線上的香港常見樹木網頁，為市民認識香港樹木提供了重要資料。本書輯錄了一百種景觀樹木品種，包括原生及引入樹種，圖文並茂，深入淺出地介紹各樹木的詳細資料，為城市植樹選種提供寶貴的信息，並同時提高讀者對樹木的鑑別能力、欣賞樂趣和保護意識。

　　張浩博士是香港園藝專業學會的現屆副會長，因此我們常有機會接觸，但我與他早於十多年前已相識。他是我的博士生，研究有關香港石礦場在各個修復階段上的植被組成和結構及土壤變化，在那時已感受到他對植物的喜愛及熱誠。近年，他一直致力在各個層面推動園藝和樹藝在香港的發展，包括促進本地從業員的專業發展、推動本地學術及教育工作，以及提升大眾對有關課題的認識。

　　作為一直關心本地綠化發展的一員，實在欣喜看見張博士繼撰寫《香港常見地被植物》後再次出版有關香港常見植物的書籍；今次的書中主角，由地被植物換成了喬木（書中另提及少量灌木）。本書包含了一百種近年香港常見的景觀樹種，配以彩色圖片，詳細介紹它們的學名科名、原產地、本地分佈狀況、生長習性、辨認特徵（高度、樹皮、葉、花、果）、花期和果期。其中一個值得推薦之處，是本書除了描述辨認特徵外，也娓娓道出各個樹種的有關趣聞（名字由來、拉丁學名含意、生態知識、用途），使讀者在認識這些樹木時，可以更添趣味。

　　我深信本圖鑑是每位園藝師、樹藝師和景觀設計師的重要工具書，樹木和植物愛好者的必然藏書，也是普羅大眾認識不同樹木時可用到的上佳參考材料。我期望本書能讓更多人發現樹木之美，愛惜這一棵棵的綠色寶藏。

朱利民　教授

香港園藝專業學會會長
香港中文大學環境、能源及可持續發展研究所院士
香港中文大學生命科學學院客座副教授
2023 年 6 月

自序

　　香港樹木存在於一個獨特的都市群落生境之中，根據植物生理生態特徵、城市發展和人類活動、地理氣候條件等相互作用下，交織出一幅「綠韻悠揚」的秀麗景觀。香港位於北回歸線以南約 130 公里，位於中國南部沿海，由香港島、九龍、新界陸地、大嶼山及其他 100 多個小島組成，天然地勢崎嶇，生態系統資源豐富。香港已建設地區僅為 24.3% 的土地面積。香港受季節性對比鮮明的亞洲季風系統影響，夏季炎熱而潮濕，冬季涼爽而乾燥，年降雨量平均超過 2300 毫米，八成降雨集中在 5 月至 9 月，同期又會受不同強度的熱帶氣旋吹襲，而且隨着近年氣候變化影響，年降雨量正慢慢上升，對都市韌性帶來不少考驗。

　　目前香港常見樹木景觀，反映了本地城市形態、景觀規劃、園藝傳統、管理系統和市民喜好等多因素下交織而出的城市森林系統。近年社會對樹木於城市生態重要性的認識愈發加深，同時亦孕育出越來越多的都市植物愛好者和研究者。隨着香港陸續出版了《香港植物誌》、《香港植物名錄》、《香港野外樹木圖鑑》等工具書，為市民認識香港植物提供了最基礎資料和重要參考。本圖鑑出版的目的，是為宣傳香港地區植物的多樣性，亦為有志於本地植物研究以及愛好者提供有用參考，並同時提高市民對樹木的識別能力和保護意識。通過實地考察搜集資料，本書共記錄了100 種景觀樹木，包括原生或由外地引入的品種，配以彩色相片，深入淺出地介紹百種樹木的詳細資料，希望各讀者於閱讀時能享受到樹木鑑別的樂趣。由於水準有限，本書如有任何疏漏之處，懇請各位讀者不吝批評指正。

　　在此衷心感謝香港特別行政區政府發展局綠化、園境及樹木管理組，對此項目的支持。同時感謝香港高等教育科技學院（THEi）、香港園境承造商協會、香港園藝專業學會，及 THEi 園藝樹藝及園境管理（榮譽）理學士課程及畢業生為這個項目所付出的努力。

　　最後，此書獻給我最敬愛的恩師莊雪影教授！

張浩 博士

香港高等教育科技學院（THEi）語文及通識教育學院學院主任

園藝樹藝及園境管理（榮譽）理學士課程主任

2023 年 6 月

目　錄

* 本書樹木學名主要根據香港植物標本室網頁。
Scientific names of trees in this book are basically
according to the Hong Kong Herbarium website.

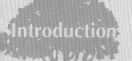
導讀：樹木的分類和術語

樹木被認為是現代城市中不可缺少的綠色基礎建設主要成分之一，亦是城市生態系統中極重要的自然組成部分。景觀樹木是泛指適於城市園林綠地及風景區栽植應用的木本植物，在莖、葉、花、果、樹形方面具有較高的觀賞價值。在人煙稠密的大都會，景觀樹木提供多樣的生態系統服務，並大大地豐富了城市的生物樣性。景觀樹木為都市可持續發展貢獻良多，在經濟、社會以及健康等多方面帶來珍貴的效益。

一、植物自然分類和人為分類

植物分類有助於我們認識和利用植物，而自然分類系統和人為分類系統是兩種主要的植物分類方法。

自然分類法依據植物的親緣和進化關係進行分類，着重反映植物物種間親緣和演化關係的親疏。一些較常採用的自然分類系統包括恩格勒（Engler）系統、哈欽松（Hutchinson）被子植物分類系統及克朗奎斯特（Cronquist）被子植物分類系統等。*Flora of Hong Kong*（《香港植物誌》）的裸子植物分類則採用了 1990 年 Kubitzki 的分類系統，被子植物分類就採用 1988 年 Cronquist 的分類系統；本書的植物分類，也是按照 *Flora of Hong Kong* 所採用的系統。

種子植物可分為裸子植物和被子植物兩大類。裸子植物是原始的種子植物，胚珠裸露，樹木中屬裸子植物的例子有馬尾松、側柏、南洋杉、落羽杉等。被子植物又稱為有花植物，與裸子植物的主要不同在於被子植物的胚珠包被於子房中，在植物授粉受精後，子房會發育成為果實，胚珠會發育為種子。另外，被子植物具有真正的花。被子植物再可分為雙子葉植物和單子葉植物。

人為分類法是人類因應不同目的，根據植物的不同特點，如生長習性、園林用途、觀賞特性等，主觀地把植物劃分成不同的大類。以景觀樹木為例，如按生長習性分類，可分為喬木類、灌木類、藤蔓類；如按園林用途分類，可分為行道樹、園景樹、庭蔭樹、綠籬樹、盆景樹等；如按觀賞特性分類，可分為觀葉類、觀花類、觀果類等。以下就簡單介紹如何按觀賞特性為景觀樹木作分類，並以本書中所提到的品種作例子：

1. 觀葉樹木

指其葉片顏色、形狀、大小或排列方式有獨特不同之處的樹木，如花葉垂葉榕、黃榕、嶺南槭等；

2. 觀形樹木

指其樹冠的形態具較高觀賞價值的樹木，如南洋杉、南洋楹、檳榔等；

3. 觀花樹木

指其花擁有出眾的花色、花形或花香的樹木，如含笑、鐘花櫻桃、絲木棉、藍花楹等；

4. 觀果樹木

指擁有明顯和豐滿果實，且掛果時間長的一類樹木，如桑、貓尾木、草莓番石榴等；

5. 觀枝幹樹木

指其枝、幹具獨特形態、色澤或附屬物等的一類樹木，如聚果榕、絲木棉、棕櫚等。

二、樹木形態學的術語

樹木的形態特徵幫助我們辨認樹木，而利用樹木形態術語，可以更準確的描述和鑑別樹木。

樹形是由樹冠及樹幹所構成的樹木形狀；不同的樹木具有不同樹形，主要由遺傳而決定，但也受外界環境因素影響。一般景觀樹木的樹形有圓柱形、尖塔形、圓錐形、卵形、倒鐘形、扁球形、傘形、垂枝形等。

樹皮的特徵包括光滑、橫紋、縱裂、鱗片狀等；樹皮的特徵隨着樹木的成長或會有所改變。

樹木的葉形變化萬千，即使是同一樹種的葉片，也有機會出現不同形態。葉形的例子包括線形、披針形、橢圓形、卵形、圓形、倒卵形、鐮刀形、三角形和盾形等。另外，葉尖、葉基、葉緣和葉脈式的特徵也會被用來描述葉片。還有，葉子可以分為單葉和複葉兩大類，而複葉又可再細分為羽狀複葉（奇數、偶數、二回）、掌狀複葉、三出複葉（羽狀、掌狀）和單身複葉。不同的葉片特徵，帶來不同的觀賞特性，例如旅人蕉、棕櫚、海棗等擁有具大葉片的樹木，就令人感受到熱帶情調。

樹木的花基本由花萼、花冠、雄蕊和雌蕊組成，可分為裸花、單被花和雙被花。花序是指花在花枝上的排列情況，有總狀花序、穗狀花序、肉穗花序、柔荑花序、傘形花序、傘房花序、複傘形花序、頭狀花序、隱頭花序等。

樹木的果實由一朵或多朵花中的子房或心皮形成，可分為單果、聚合果和聚花果。單果按照果皮及其附屬部分於成熟時的質地和結構，可分為肉質果和乾果。肉質果的類型有漿果、瓠果、核果、梨果、柑果；乾果的類型有蓇葖果、莢果、蒴果、長角果、穎果、瘦果、堅果和翅果等。

本圖鑑按照樹木的分科排序，詳細介紹樹木的學名、英文名、中文名、科名、本地分佈狀態、原產地、植物知識與趣聞（名字由來、應用、生態）、生長習性、高度、辨認特徵（樹幹、樹皮、葉、花、果）；讀者在觀樹時，可以按着圖鑑中提出的辨認特徵識別樹木。

Guide: Classification and Terminology of Ornamental Trees

Trees are considered to be one of the main components of the indispensable green infrastructure in modern cities, and also an extremely important natural part of the urban ecosystem. Ornamental trees generally refer to woody plants that are suitable for planting in urban gardens and scenic spots, and have high ornamental value in terms of stems, leaves, flowers, fruits, and tree shapes. In densely populated metropolises, ornamental trees provide diverse ecosystem services and greatly enrich urban biodiversity. These trees greatly contribute to the sustainable development of cities and bring valuable benefits in economic, social and health aspects.

A. Natural classification and artificial classification of plants

The classification of plants helps us to understand and utilize plants. Natural classification and artificial classification are the two main approaches to classifying plants.

Natural classification is the classification of plants based on their phylogenetic and evolutionary relationships. It shows the phylogenetic and evolutionary affinity among the different plant species. The Engler System, the Hutchinson System of Classification of Angiosperms, and the Cronquist System of Classification of Angiosperms are some of the commonly adopted classification systems. The *Flora of Hong Kong* adopted the classification system of Kubitzki in 1990 for the classification of gymnosperms and that of Cronquist in 1988 for the classification of angiosperms. The plant classification in our book followed the systems adopted by the *Flora of Hong Kong*.

All seed plants are divided into two categories, angiosperms and gymnosperms. Gymnosperms are primitive seed plants and have naked ovules. Some examples are Chinese Red Pine (*Pinus massoniana*), Chinese Arborvitae (*Platycladus orientalis*), Hoop Pine (*Araucaria cunninghamii*) and Bald Cypress (*Taxodium distichum*). Angiosperms are also known as flowering plants. Their main differences with gymnosperms are that their ovules are enclosed within the ovary. After pollination and fertilization, the ovary will develop into a fruit, and the ovules will develop into seeds. In addition, angiosperms produce true flowers. The angiosperms can be further classified into dicots and monocots.

The artificial classification method is that humans subjectively classify plants into different categories according to different characteristics of plants, such as growth habit, garden uses and ornamental characteristics, for different purposes.

Taking ornamental trees as an example, if they are classified according to their growth habits, they can be divided into trees, shrubs and vines; if they are classified according to their garden uses, they can be divided into street trees, landscape trees, shade trees, hedge trees and bonsai trees, etc.; according to the classification of ornamental characteristics, it can be divided into ornamental foliage trees, ornamental flowering trees, ornamental fruit trees and so on. The following is a brief introduction to how to classify landscape trees according to their ornamental characteristics, and use the species mentioned in this book as examples:

1. Ornamental foliage trees

The trees with unique differences in leaf color, shape, size, or arrangement, such as Variegated Weeping Fig (*Ficus benjamina* 'Variegata'), Golden Chinese Banyan (*Ficus microcarpa* 'Golden Leaf') and Tutcher's Maple (*Acer tutcheri*).

2. Ornamental structural trees

Trees with crown shape or structure that is of high ornamental value, such as Hoop Pine (*Araucaria cunninghamii*), White Albizia (*Falcataria moluccana*) and Betel Palm (*Areca catechu*).

3. Ornamental flowering trees

Trees with flowers having outstanding flower colour, shape or fragrance, such as Figo Michelia (*Michelia figo*), Bell-flowers Cherry (*Prunus campanulata*), Silk Floss Tree (*Chorisia speciosa*) and Jacaranda (*Jacaranda mimosifolia*).

4. Ornamental fruit trees

Trees with prominent and abundant fruits, and with a long fruiting period, such as White Mulberry (*Morus alba*), Cat-tail Tree (*Markhamia stipulata*) and Strawberry Guava (*Psidium cattleianum*).

5. Ornamental stem trees

Trees with branches and trunk having unique forms, colours, or objects on them, such as Cluster Fig (*Ficus racemosa*), Silk Floss Tree (*Chorisia speciosa*) and Windmill Palm (*Trachycarpus fortunei*).

B. Terminology of tree morphology

The morphological features of trees help us to identify a tree. Using the terminology of tree morphology allows us to describe and identify a tree more accurately.

Tree form is the shape of a tree, formed by the tree crown and main stem. Different trees have different tree forms. The tree form is mainly determined by genetic factors, but also affected by environmental factors. Some common tree forms of ornamental trees include columnar, pyramidal, conical, oval, vase, round, spreading, weeping, etc.

The characteristics of tree bark include smooth, horizontal stripes, longitudinal fissures, scales etc. The characteristics of tree bark may change as the tree grows.

The shapes of tree leaves are highly varied. The leaves of the same tree species may have different shapes. Some examples of leaf shapes include linear, lanceolate, elliptic, ovate, orbicular, obovate, falcate, deltoid, and peltate. The characteristics of leaf apices, leaf base, leaf margins, and leaf venation are also used to describe a leaf. In addition, leaves are divided into single leaves and compound leaves. Compound leaves are further classified into pinnate (odd-pinnate, even-pinnate, bipinnate), palmately compound, trifoliolate (pinnately, palmately) and unifoliate compound. Leaves with different characteristics are with different ornamental features. For example, tree species with giant leaves, such as Traveller's Palm, Windmill Palm and Date Palm, give a tropical aura.

The tree flowers are basically composed of calyx, corolla, androecium and gynoecium, and are classified as naked flowers, chlamydeous flowers, and dichlamydeous flowers. Inflorescence is the arrangement of flowers on an axis. The inflorescence types include raceme, spike, spadix, catkin, umbel, corymb, compound umbel, capitulum, hypanthodium, etc.

The tree fruits are developed from the ovaries or carpels from one or more flowers. The fruit types include simple fruits, aggregate fruits and multiple fruits. Based on the nature of the pericarp and associated accessory structures when they are ripened, the simple fruits are classified into fleshy fruits and dry fruits. The types of fleshy fruits include berry, pepo, drupe, pome, hesperidium, and the types of dry fruits include follicle, legume, capsule, silique, caryopsis, achene, nut, samara, etc.

This photographic guide is sorted by tree family, and introduces in detail the scientific name, English name, Chinese name, family name, distribution, origin, species knowledge and anecdotes (origin of the name, application and ecology), growing habit, height, and traits for identification (trunk, bark, leaves, flowers, fruits) of the trees. Readers can identify trees according to the identification features listed in the photographic guide when appreciating trees.

香港市區樹木栽種的歷史與轉變

戰後「先鋒樹種」固定水土

二次世界大戰期間，香港經歷戰火的洗禮，大多數市區建築物和基礎設施在淪陷期間遭到破壞，郊野地區光禿處處。政府早期的植林工作，主要以固定斜坡和美化市容為目的。考慮到當時需要種植的土地貧瘠及面積廣闊，由二戰後直至 70 年代，均以大量引入適應力強，適合本地環境，且生長速度較快的「先鋒樹種」為主，例如濕地松、台灣相思、白千層、紅膠木等。政府亦於市區廣泛種植生長速度快且樹形優美的品種，例如鳳凰木、南洋杉、假檳榔和秋楓等。

增原生樹木
成香港獨特林木景觀

80 年代開始，政府開始着重發展康樂地區、樹木教育區和修補曾受破壞的景觀區，大量種植本地原生品種，例如楓香、樟樹、荷木、黧蒴錐等；另外增補代表性的樹木，例如潤楠屬、殼斗科、山茶科、杜英科、金縷梅科等樹木；建設「標本林」及在適當地方種植季節色彩的品種，例如位於大棠的楓香林。此外，對於水土沖刷嚴重的地區，採用多種原生及外來品種，實行「混植」並轉為種植「混交林」；曾大規模種植「先鋒樹種」的「純林地區」，則進行間植原生品種，例如種植本地的闊葉樹種。在長遠的歲月中，原生和外來樹種相互交織，形成了香港城市林木獨特的景觀。

優化管理規劃 改善樹木問題

香港早期大量種植的外來樹種生長快速，能迅速覆蓋郊野、道旁及休憩場所的裸露的土地，並具有一定觀賞價值，惟外來樹種不能與本地原生動植物構成良好生態關係，部分外地樹種更有一定的排他性，故對本地生態系統貢獻有限。當時種植的樹木，雖然在短時間內緩解了水土流失，並有固定斜坡及綠化環境之效，但樹木的生命有一定週期，當時種植的樹木在多年後開始出現老化或壞死。

香港城市發展迅速，樹木在市區道路旁的生長空間愈發受限，不少樹木的樹根生長空間嚴重不足，無法可持續地支撐樹木的生長。香港位處亞熱帶地區，天氣潮濕炎熱，極端天氣愈發常見，細菌、真菌、害蟲在此氣候條件下對樹木生命及成長造成極大威脅。綜合以上種種因素，香港樹木種植和管理的形勢愈加複雜；隨着市民對生態和環保意識提高，更全面和長遠的管理規劃及政策發展開始受到社會重視。

考慮到上述的背景和原因，香港在過去數十年間，對於栽種樹木的目的和策略，亦有着明顯的轉變。除了景觀綠化及固定水土之功能外，政府開始重視

樹木對於保育價值和生物多樣性的作用，推行更多元化和可持續的樹木管理策略，逐步改變了樹木管理及栽種樹木的方式和目標。2009年底，政府開始了「植林優化計劃」，目的是提高可持續性、生物多樣性，改善林區生態及樹木健康，減少病蟲害。推行方式是在外來品種較多的老化林地進行疏伐，並補種本地原生種的樹苗，加強管理，使林區的生態環境改變，把原本的單一「先鋒樹種」的樹林逐步優化，改變成原生種的次生林。

成立專責小組
倡「植樹有方，因地制宜」

由於香港樹木分佈極廣，政府認為若以單一部門處理樹木的管理工作是不理想及不可行的，故採取綜合策略管理，釐清各部門的分工，按其所在位置歸屬不同政府部門管理，提升綠化計劃的協調能力及更有效執行樹木管理工作。發展局在2010年3月成立綠化、園境及樹木管理組，每個部門配有專業人員負責樹木的管理和種植，肩負綠化、園境及樹木管理的整體政策責任，通盤地進行園境及樹木管理工作。

各樹木管理部門則根據管理組提出「植樹有方，因地制宜」的大原則，考慮種植目標、地點及空間限制、周邊環境、微氣候等因素後，選擇合適的植物品種在適當的地點種植，以確保植物能持續生長，有利長遠生態發展。在此原則下，政府各部門不再只是隨意地在各處種植外來樹種，而是根據不同地區的特色和需要，選擇合適的本地或原生樹種，整體地提升城市林木的健康、適應力和多樣性。

The History and Changes of Tree Planting in Hong Kong Urban Areas

Introduced "Pioneer Tree Species" after the Baptism of War

During the Second World War, Hong Kong experienced the baptism of war. Most of the urban buildings and infrastructure were destroyed during the occupation, and the rural areas were left bare. The government's early afforestation work was mainly aimed at fixing slopes and improving the city's overall greening appearance. Considering the barren and vast area of land that needed to be planted at that time, from the end of World War II until the 1970s, a large number of "pioneer tree species" were introduced, such as Slash Pine (*Pinus elliottii*), Taiwan Acacia (*Acacia confusa*), Paper-bark Tree (*Melaleuca cajuputi* subsp. *cumingiana*), and Brisbane Box (*Lophostemon confertus*), etc. These exotic species are suitable to plant in the local environment as of their strong adaptability and fast growth rate. The government has also widely planted some fast-growing and beautiful tree varieties in urban areas, such as Flame Tree *(Delonix regia)*, Hoop Pine (*Araucaria cunninghamii*), Alexandra Palm (*Archontophoenix alexandrae*) and Autumn Maple (*Bischofia javanica*), etc.

Increased Native Trees Forming a Unique Forest Landscape in Hong Kong

Starting from the 1980s, the government shifted its focus on the development of recreational areas and tree education areas and the restoration of damaged landscapes. A large number of local native species, such as Sweet Gum (*Liquidambar formosana*), Camphor Tree (*Cinnamomum camphora*), Schima (*Schima superba*), Chestnut Oak (*Castanopsis fissa*), etc., were planted. In addition, some representative native species, such as *Machilus spp.*, Fagaceae, Theaceae, Elaeocarpaceae, Hamamelidaceae, etc., were also planted. Meanwhile, the "specimen forests" were established and some species with seasonal colours in appropriate places were planted, for example Sweet Gum (*Liquidambar formosana*) forest in Tai Tong. Besides the above afforestation efforts, for areas with severe soil and water erosion, a variety of native and exotic species were adopted and "mixed-planted" was implemented and converted to mixed forests. For "pure plantation areas" where "pioneer species" had been planted on a large scale, native species such as broad-leaved tree species were interplanted. Throughout the years, exotic and native tree species have grown together and intertwined into the unique urban forest landscape in Hong Kong.

Growing Complexity of Tree Problems Urged Better Management of Tree Planting

The exotic tree species planted in Hong Kong in the early days have a fast growth rate, and they quickly covered the bare land in the countryside, roadsides and rest areas. It is beyond doubt that these exotic species have a certain ornamental value. However, they cannot form mutually inclusive ecological relationships with the native flora and fauna, and some of them are even highly competitive in biological exclusion. Therefore, their contribution to the local ecosystem is limited. The vast planting of these trees alleviated the problem of soil erosion and contributed to slope stabilisation and environment greening in a relatively short period. However, these trees have a finite life cycle. Many of these trees started ageing and becoming necrotic after many years.

Amid the rapid urban development of Hong Kong, trees beside urban roads have been struggling as growth space has become more and more limited. Many trees have insufficient space for their roots to grab ground soil tightly, and thus further growth is not supported. As a subtropical zone with hot and humid weather, extreme weather is also becoming common in Hong Kong. Bacteria, fungi, and pests pose a serious threat to the health of the trees under such climatic conditions. Combining all the above factors, the situation of tree planting and its management has become more and more complicated in Hong Kong.

As the public's awareness of ecology and environmental protection has increased gradually, society is now placing great value on more comprehensive and long-term management planning and policy development.

Considering the background and reasons above, the Hong Kong government has undergone evident changes in the purpose and strategies of tree planting in the past few decades. In addition to the functions of landscape greening and water and soil stabilisation, the government has paid more attention to the role and value of trees in conservation and biodiversity. It has implemented more diversified and sustainable tree management strategies, by which gradually changing the approaches and purposes of tree management and planting trees. At the end of 2009, the government launched the "Afforestation Optimisation Programme", which aims to improve sustainability, biodiversity, forest ecology and tree health, as well as to mitigate pests and diseases of local forestry. The implementation method was to carry out thinning in ageing forest land with many exotic species, and replant saplings of local native species. Moreover, the programme has also strengthened tree management, improved the forestry ecology, and gradually optimised the mono-cultural "pioneer tree species" forests and changed them into secondary forests of native species.

Established Task Force Following the Principle of "Right Plant, Right Place"

Due to the wide distribution of trees in Hong Kong, the government believes that it is unsatisfactory and unfeasible for a single department to deal with the management of trees. Therefore, it has adopted an approach of comprehensive strategic management by first clarifying the accountabilities of diverse government departments in tree management, where responsibilities belong to individual departments according to the trees' locations. Thus, boosting the coordination of greening programmes and improving the efficiency of tree management. In March 2010, the Development Bureau established the Greening, Landscape and Tree Management Section. Each department has been assigned professionals for the management and planting of trees. They shoulder the overall policy responsibility for greening, landscape and tree management, and carry out the work in a holistic manner. Those tree management departments, dispersed in the government organisation, started following the principle of "Right Plant, Right Place", which was strongly advocated by the Section. They select suitable plants for planting in appropriate places with due consideration of various factors such as the planting objective, site and spatial constraints, surrounding landscape character, microclimate, etc., to ensure sustainable plant growth and benefit the ecological development in the long term. Under the principle, various government departments no longer just randomly plant exotic tree species everywhere, but choose appropriate local or native tree species according to the characteristics and needs of different regions, so as to enhance the health, resilience and diversity of the urban forests as a whole ecosystem.

二維條碼樹木標籤

　　全港樹木分佈極廣，由政府負責恆常護養的樹木當中，約有超過一半以上位於人流車流密集的地點。要令樹木健康成長，除了有賴政府部門的綜合管理和日常護養之外，市民的積極參與亦十分重要。政府發展局一直積極探索如何應用智慧科技，來提升樹木管理的成效。為加強市民對樹木的認識和關注，二維條碼樹木標籤就是其中一項近年推行的新措施。

　　發展局於 2020 年 1 月開始展開二維條碼樹木標籤工作，項目於 2022 年初首先為約 20 萬棵於行人路旁、公園、花園、休憩處及公共屋邨的樹木安裝二維條碼樹木標籤。除了發展局樹木登記冊內的樹木外，被揀選的樹木需生長於行人可到達的位置，並位於智能手提電話閱讀二維條碼的掃瞄距離範圍內。二維碼標籤的樹牌上印有樹木的基本資料，包括其中英文名及學名，而其懸掛高度亦需配合行人視線，務求讓市民能輕易發現並使用這些標籤。

　　透過掃瞄二維碼標籤，市民可以輕鬆地以智能手提電話查閱該樹種的詳盡資訊，例如樹木品種，屬於原生還是外來物種、樹木特徵、植物趣聞等，從多角度了解身邊的樹木品種，以加深市民對樹木的愛護之情。

　　此外，二維條碼樹木標籤亦為市民報告問題樹木提供便利，由於樹牌上印有每棵樹的獨有編號，故能有助市民準確說出樹木位置，只要透過「1823」一站式系統，即可以電話、電郵、網站或於應用程式查詢或報告標有樹木編號的問題樹木，令整個程序變得準確和輕鬆。

　　樹木在城市的可持續發展中佔有越來越重要的地位，它們不僅為社區景觀添上自然綠意，更能緩和氣溫、改善空氣質素，並提升生物多樣性。二維碼樹木標籤計劃成功把樹木護養帶入市民日常生活之中，透過培養公眾對身邊一樹一木的了解和尊重，和方便市民參與護樹工作，必能使樹木更茁壯成長，讓香港成為一個宜居城市。

Tree Labels with QR Codes

The distribution of trees is wide in Hong Kong. Among the trees under the management of the government, more than half of them are in places with heavy traffic flows of pedestrians and vehicles. Apart from the government departments' comprehensive management and regular maintenance, public participation is indispensable to ensure the healthy growth of trees. The Development Bureau has been actively exploring smart technology applications to improve tree management efficiency. To engage the public, the "Tree Labels with QR Codes" is one of the new measures implemented in recent years to enhance public awareness and concerns about the maintenance of trees.

The Development Bureau commenced the project in January 2020. Since early 2022, labels with QR codes have been displayed on about 200,000 trees along sidewalks and roadsides or those in parks, recreational areas and public housing. The tree selected have to be registered under the Development Bureau and accessible to pedestrians within the scanning distance of a smartphone. Aside from the QR code and Tree Number, these labels also show basic information about the trees, including their Chinese and English names, and their botanical names. The labels must be displayed at the eye level of pedestrians whereby the public could easily spot and access them.

By simply scanning the QR code, the public can get much information about the tree, such as tree species, whether of native or exotic origin, characteristics, anecdotes, etc. The ready-to-access information allows ordinary citizens to understand the tree species in multi aspects. It also helps to deepen public appreciation of these woody plants.

In addition, these tree labels make reporting problematic trees more convenient. The public could accurately report the location of any single tree by quoting its unique Tree Number printed on the label. Through the 1823 system, one can inquire about or report problematic trees marked with the Tree Number by phone, email, website or app. Reporting problematic trees is never easier and more accurate.

Trees play an increasingly significant role in the sustainable development of our city. They not only add natural greenery to our community landscapes but also moderate temperatures, improve air quality, and promote the biodiversity of our habitats. The "Tree Labels with QR Codes" has successfully brought tree maintenance and assessment into the public's daily life. By cultivating knowledge and respect for trees among the public and making participation in tree conservation more convenient, our urban forest will surely grow stronger and make Hong Kong becoming a more livable city.

「1823」24 小時電話熱線：1823

電郵地址：tellme@1823.gov.hk

網站：https://www.1823.gov.hk/tc

手機應用程式：Tell me@1823 v2

觀樹及樹木攝影守則

為了減少觀樹及樹木攝影過程中對樹木的干擾或傷害，筆者制訂了一套守則供市民參考，旨在教導市民如何欣賞和保護樹木，希望可以作為一套觀樹及樹木攝影活動的良好行為規範。

1. 愛護樹木

無論是進行觀賞或攝影活動時，應盡量以不影響自然環境及不損害樹木為原則，以免造成干擾或傷害，

a. 不要切割、採摘或根除任何樹木或樹木的任何部分，除非獲得合法許可證；

b. 不要嘗試吞吃樹木任何一部分，不要觸摸未知樹木品種的任何部位；

c. 不要攀爬或折斷樹枝，不要刻畫或塗污樹幹，不要挖掘或填埋樹根，不要在樹上掛掛鈎或其他物件；

d. 不要用火或煙對樹木造成傷害，不要用化學物質或有害物質污染樹木或其周圍的土壤和水源，不要用尖銳或硬物敲打或戳刺樹木；

e. 不要干擾樹上鳥巢或蜂巢，以免親鳥棄巢或招來蜂類攻擊；及

f. 不要隨意破壞或改變樹木的自然生長環境，如移走岩石、枯枝、落葉等。

2. 報告有問題樹木

在觀樹及樹木攝影時，如發現樹木不尋常情況，例如：

- 死樹
- 傾斜
- 樹枝枯死
- 懸吊斷枝
- 枝幹／V 型樹椏有裂縫或裂開
- 等勢莖
  （即樹幹的接合點出現裂縫或腐爛）
- 樹幹出現腐爛或樹洞
- 樹根被嚴重切割或損害
- 呈現真菌子實體
- 病蟲害

應加倍留意，並盡快致電 1823 或填妥網上表格 (https://www.1823.gov.hk/tc/form/complain/tree)，以便相關部門跟進有關問題樹木。

3. 舉報干擾及傷害樹木不尋常行為

如果發現有人干擾或傷害樹木，在安全情況下宜向他們解釋和勸止。如果未能阻止，請拍照記錄，並盡快向漁農自然護理署或致電 1823 舉報。

4. 尊重他人

a. 避免干擾其他遊客，讓大家都可以享受其中的樂趣；

b. 小心不要破壞周邊的設施與環境。

觀樹是一項欣賞自然美景的活動，也是一種提升環保意識的方式。遵守觀樹守則，讓人們能夠在欣賞和學習樹木的同時，尊重和保護樹木的生命和權利，並與其他生物和諧共處，傳達對自然的尊重和愛護。

Code of Conduct for
Tree Appreciation and Tree Photography

To reduce the disturbance or harm to trees during tree appreciation and tree photography, Dr Allen ZHANG has drawn up the following codes as a reference, aiming to educate citizens how to appreciate and conserve trees, and hope it can provide a model for good practices in tree appreciation and tree photography.

1. Conserve tree

Tree appreciation and tree photography should be carried out with minimum interference to trees and their environments. Disturbance or harm must be avoided as far as possible,

a. Do not cut, pick, uproot any tree or part of a tree, except in accordance with a legal permit;

b. Do not try to eat any part of a tree, and do not touch any part of a tree for unknown species;

c. Do not climb or break branches, do not carve or deface tree trunk, do not dig or bury tree roots, and do not use hooks or hook other objects on trees;

d. Do not cause harm to trees with fire or smoke, do not contaminate trees or the surrounding soil and water with chemicals or harmful substances, and do not hit or poke trees with sharp or hard objects;

e. Do not disturb the bird nest or beehive in the tree, lest the parent birds abandon the nest or attract bee attack;

f. Do not destroy or alter the natural growth environment of tree, such as removing rocks, dead branches and fallen leaves.

2. Report unusual tree

During the tree appreciation and tree photography, if there are any abnormalities, for example:

• Dead tree

• Leaning
• Dieback twig
• Hanger
• Branch / V-shaped crotch with cracks or split
• Codominant stems (i.e. the weak union with crack and/or decay on the trunk)
• Wood decay or cavity in the trunk
• Severely cut or damaged root
• Fungal fruiting bodies
• Pest problem

Please stay vigilant and call 1823 or complete the online form (https://www.1823. gov.hk/tc/form/complain/tree), to facilitate the responsible department to take action as soon as possible.

3.Report disturbance and harm to trees

If there are any disturbing or causing harm to trees, advise against the act when it is safe to do so. If fail to stop it, please take photos and report to the Agriculture, Fisheries and Conservation Department or report by giving a phone call to 1823 as soon as possible.

4. Respect others

a. Share the fun and avoid disturbing other visitors on site;

b. Take care not to damage the public facilities and the environment.

Tree appreciation is an activity to appreciate the beauty of nature and a way to raise awareness of environmental protection. Observing the code of conduct for tree watching and tree photography allows people to appreciate and learn from trees, respect and protect the lives and rights of trees, and live in harmony with nature, conveying respect and love for nature.

香港 100 種
景觀樹木

100 ORNAMENTAL TREES of HONG KONG

南洋杉

Hoop Pine | *Araucaria cunninghamii* Aiton ex D. Don

相片拍攝地點：香港中文大學神學院
Tree Location: Divinity School of Chung Chi College, The Chinese University of Hong Kong

名字由來 MEANINGS OF NAME

屬名 *Araucaria* 意指位於智利的阿勞科省，該省正是首次發現智利南洋杉的地方。為紀念「澳洲兄弟」——艾倫·坎寧安及理查德·坎寧安（1791-1839 及 1793-1835），種加詞*則取自其姓氏 *cunninghamii*。他們皆是邱園（英國皇家植物園）的傑出植物學家，同時因探索澳洲東部而聞名於世。

*種加詞（specific epithet），又稱「種小名」、「種本名」，是雙名法中物種名的第二部分。雙名法的第一部分為屬名，第二部分為種加詞，常為形容詞，用來修飾屬名並解釋此品種的性質。

The generic name *Araucaria* refers to Arauco, a province of Chile where the first *A. araucana* (Chilean pine) was observed; Chilean pine was the first published plant in this genus. The specific epithet *cunninghamii* is in honour of the "Australian brothers" Allan and Richard Cunningham (1791-1839 and 1793-1835). They were eminent botanists in Kew herbarium and well-known with their exploration of eastern Australia.

本地分佈狀態 DISTRIBUTIONS	外來物種 Exotic species
原產地 ORIGIN	分佈狹窄，原生於新幾內亞和澳洲東部，如新南威爾士州和昆士蘭州。 Narrow range in New Guinea and eastern Australia, including New South Wales and Queensland.
生長習性 GROWING HABIT	常綠喬木。於香港高度達 30 米，於原產地可高達 70 米。金字塔形樹冠。 Evergreen tree. Up to 30 m in Hong Kong, and up to 70 m in its place of origin. Pyramidal crown.

花果期月份	1	2	3	4	5	6	7	8	9	10	11	12

花期：本港二月至六月。果期：本港六月至八月。
Flowering period: February to June in Hong Kong. Fruiting period: June to August in Hong Kong.

① 樹幹 TRUNK	② 樹皮 BARK	③ 葉 LEAVES
④ 孢子葉球 POLLEN CONE	⑤ 毬果 SEED CONE	

① 南洋杉的樹幹。

Trunk of *Araucaria cunninghamii* Aiton ex D. Don.

② 樹皮暗灰色或灰棕色，橫狀裂開，粗糙。

Bark dark greyish or greyish brown, transversely split, coarse.

③ 葉片簇生於側枝頂端，二形性。幼樹或成熟後下部枝條的葉片呈針形或鐮形。成熟後上部枝條或具毬果的枝條葉片呈卵形或三角狀卵形，葉片密集。葉面具明顯氣孔線，呈白色。

Leaves clustered at the apex of lateral branchlets, dimorphic. Leaves on young trees or lower branchlets of mature trees acicular or falcate. Leaves on upper or cone-bearing branchlets of mature trees ovate or triangular-ovate, densely arranged. Obvious stomatal lines adaxially, white.

④ 頂生。雄毬果圓柱形。雌毬果球形至卵毬形，比雄球果大，呈綠色。

Terminal. Male cones cylindrical. Female cones globose to ovoid, relatively larger, green.

⑤ 毬果球形至卵球形，成熟時由綠色轉為褐色。種子橢圓形，具珠鱗，由側芽發育而成。

Seed cones globose to ovoid, turning from green to brown when mature. Seeds ellipsoid, with ovuliferous scale, developed from a modified lateral shoot system.

應用 APPLICATION

南洋杉柔軟，不太耐用，為滿足製作家具、模具及其他輕型建材等大量木材需求，澳洲大量砍伐南洋杉以取得其木材。值得注意的是，於 2008 年，南洋杉植被在昆士蘭州的森林擴展了 26%，反觀野生南洋杉的數量因進化和人為壓力等因素而受到限制。隨着種群數量的不斷減少，種群更容易受到瓶頸效應的影響，從而導致種群遺傳多樣性急劇下降。因此迫切需要進行相關研究和保育工作，以保育野生南洋杉的數量。

In respect of its soft and less-durable wood, the wood is heavily harvested in Australia to attain the demands of furniture, moulding and other light construction. In contrast to the plantation of which has expanded to 26% in Queensland's forests, the wild population of Hoop Pine is getting restricted because of evolutionary and anthropogenic pressures. When the population keeps shrinking, it is more susceptible to bottleneck effect and genetic diversity could be dramatically reduced afterwards. As a result, more studies and conservation work are urgently required to conserve the wild Hoop Pine.

植物趣聞 ANECDOTE ON PLANTS

鑑別南洋杉科樹木 Identification of Trees in Araucariaceae：

與被子植物不同，裸子植物沒有迷人的花朵和果實。取而代之的是，孢子葉球包裹着配子，種子裸露，並沒有果皮包裹着。加上形態上難以區分的針葉，故要鑑別它們困難重重。

南洋杉科樹種擁有高大且垂直的樹身，故香港常引入並廣泛種植作觀賞用途。南洋杉、異葉南洋杉和柱狀南洋杉是香港主要引入的南洋杉科樹種。成熟的南洋杉科樹種，葉片呈二形性。南洋杉的葉片多刺且鉤狀凸出，其他兩種的質感較柔和。

Unlike angiosperms, gymnosperms cannot produce glamorous flowers and fruits. Instead, their gametes are packed in pollen cones and the seeds are naked without the enclosure of pericarp. Coupled with their outwardly indistinguishable needle-like leaves, the identification work is thwarted and challengeable.

Araucariaceae are broadly introduced in Hong Kong as ornamental trees by virtue of their lofty vertical dimensions. Hoop Pine, *A. heterophylla* (Norfolk Island Pine) and *A. columnaris* (Cook Pine) are the dominant introduced Araucariaceae trees in Hong Kong. The leaves of mature Araucariaceae trees are dimorphic. Those of Hoop Pine are prominent from being spiny and hook-like, while the textures of the other two are comparably gentle.

馬尾松

Chinese Red Pine | *Pinus massoniana* Lamb.

相片拍攝地點：沙田公園
Tree Location: Sha Tin Park

名字由來 MEANINGS OF NAME

種加詞 *massoniana* 意為葉呈馬尾形狀。

The specific epithet *massoniana* in Latin refers to its horse-tail like leaves.

應用 APPLICATION

馬尾松生長迅速，具適應貧瘠土壤和不理想生長環境的特質，故在南中國被廣泛用作植林中的先鋒品種。它的原木作為建築物料、家具和紙張原料上的貢獻不可或缺。馬尾松的樹皮和針葉可作為治療神經衰弱和高血壓的輔助材料。

Chinese Red Pine is a common pioneer tree in plantation forestry in southern China due to its remarkable adaptability to barren soils and drought. In view of its fast growth rate, it is an excellent timber tree, with its wood yearned for marking furniture, wood pulp and other construction. Its barks and needles are ancillary ingredients to increase the efficacy of treatments for neurasthenia and hypertension.

本地分佈狀態 DISTRIBUTIONS	原生物種 Native species
原產地 ORIGIN	遍佈於香港及南中國省份。 Hong Kong and the Southern China.
生長習性 GROWING HABIT	常綠喬木。高度可達 45 米。 Evergreen tree. Up to 45 m tall.

1	2	3	4	5	6	7	8	9	10	11	12	花果期 月份

花期：本港四月至五月。果期：本港十月至十二月。
Flowering period: April to May in Hong Kong. Fruiting period: October to December in Hong Kong.

① 樹幹 TRUNK   ② 樹皮 BARK   ③ 葉 LEAVES
④ 孢子葉球 POLLEN CONES   ⑤ 毬果 SEED CONES

① 馬尾松的樹幹。小枝黃棕色，通常每年生長一至兩輪。

Trunk of *Pinus massoniana* Lamb. Branchlets yellowish brown, always growing once to twice per year.

② 樹皮紅棕色，具不規則鱗狀剝落。

Bark reddish brown, irregularly scaly and flaking.

③ 冬芽褐色，卵狀圓柱形或圓柱形。針葉 2 針一束，甚少 3 針一束，纖細，微扭曲，葉緣具鋸齒，氣孔於葉片兩面排列成行。

Winter buds brown, ovoid-cylindric or cylindric. Needle 2 per bundle, occasionally 3, slender, slightly twisted, serrate, stomata arranged in rows on both surfaces.

④ 雌雄同株，單性花。雄性毬果簇聚於一年生小枝基部，雌性毬果生於小枝頂端。

Monoecious. Male cones crowded at the base of annual branchlets, female cones at the apex of branchlets.

⑤ 毬果懸垂，卵圓形，成熟時由綠色轉為栗棕色，種鱗近長橢圓狀倒卵形或近長方形，鱗盾菱形。種子狹卵形，具翅。

Seed cones pendulous, ovoid, turning green to chestnut when mature, seed scales suboblong-obovoid or subsquare, apophyses rhombic. Seeds narrowly ovoid, winged.

生態 ECOLOGY

相信大部分人皆聽聞過「馬尾松」這樹種，但鮮少親眼目睹其真身。在香港開埠早期，馬尾松是優勢樹種，它見證着香港的歷史變遷。在眾多松樹中，馬尾松是唯一一種原生於香港的松樹。馬尾松能適應貧瘠土壤和不理想的生長環境，所以廣泛應用於荒山造林，以減少土壤侵蝕。

在 1894 年，馬尾松受到大量馬尾松毛蟲侵襲，造成大規模落葉、樹木群落減少等影響。這問題在 50 年代，施加化學除蟲劑後得以緩解。在日本侵華及國共內戰期間（1940-1946），馬尾松因作為木材或燃料而遭過度砍伐，導致數量銳減。

戰後，林地復育計劃在香港再次起動，廣泛種植「植林三寶」，包括台灣相思、紅膠木和濕地松，導致馬尾松族群再收窄。到了 70 年代，馬尾松受到松櫛圓盾介殼蟲和松材線蟲侵襲，馬尾松在短短十年間大量死亡。由於濕地松對蟲害的耐受性較高，它慢慢地取代馬尾松，並成為香港現時的優勢松樹品種。

I believe most people have heard about Chinese Red Pine but never or seldom see the tree. Indeed, it was dominant in early Hong Kong, with remarkable vicissitudes of history. Chinese Red Pine is the only pine tree species native to Hong Kong. By virtue of its great tolerance to barren lands, it was widely adopted as one of the framework trees in the early revitalization of barren hills to attenuate soil erosion.

In 1894, Chinese Red Pine was threatened by *Dendrolimus punctatus* (Masson Pine Caterpillar). About 36 tons of caterpillars defoliated the trees and eliminated a certain part of the population. The problem was alleviated after the application of chemical pesticides in the 1950s. Nothing lasts forever. The population of Chinese Red Pine faced the second reduction during the Japanese invasion and Chinese Civil War in Hong Kong (1940–1946). Most of the trees were logged for fuels and woods.

Reforestation restarted after the wars and the dominance of Chinese Red Pine had declined due to the increasing introduction of *Acacia confusa* (Taiwan Acacia), *Lophostemon confertus* (Brisbane Box) and *Pinus elliottii* (Slash Pine), which are currently known as The Three Treasures for Afforestation in Hong Kong. Until the 1970s, the invasions of *Hemiberlesia pitysophila* (Pine Needle Scale) and the *Bursaphelenchus xylophilus* (Pinewood Nematode) resulted in the massive deaths of Chinese Red Pine within 10 years. Since Slash Pine is less susceptible to the pests, it has substituted the Chinese Red Pine and become the dominant pine species in Hong Kong.

落羽杉

Bald Cypress, Deciduous Cypress | *Taxodium distichum* (L.) Rich.

相片拍攝地點：九龍公園、香港動植物公園、沙田公園
Tree Location: Kowloon Park, Hong Kong Zoological and Botanical Gardens, Sha Tin Park

名字由來 MEANINGS OF NAME

種加詞 *distichum* 意謂其葉在一年生小枝上排成 2 列。在冬季落葉時，它的葉片隨風散落，猶如羽毛在風中起舞，故又名落羽杉。

The specific epithet *distichous*, referring to its leaves beautifully arranged in 2 ranks on annual branchlets. On account of its leaves peeling off and driving wind like feathers in winter, it is also named as「落羽杉」in Chinese.

應用 APPLICATION

落羽杉的木材具有高度防腐性，故被廣泛用作建築物料、家具原料和船隻建材，而樹脂則具有藥用價值。鑑於其引人入勝的樹型和葉片顏色，故目前此樹種在公園和花園中被廣泛種植。

The wood of *Taxodium distichum* is highly antiseptic, so it is widely used as building materials, furniture raw materials and shipbuilding, and resin has medicinal value. Due to its attractive shape and foliage color, the tree is now widely planted in parks and gardens.

本地分佈狀態 DISTRIBUTIONS	外來物種 Exotic species
原產地 ORIGIN	北美東南部。 Southeastern United States.
生長習性 GROWING HABIT	落葉喬木。高度可達 50 米。 Deciduous tree. Up to 50m tall.

花果期 月份	1	2	3	4	5	6	7	8	9	10	11	12

花期：本港三月至四月。果期：本港七月至十月。
Flowering period: March to April in Hong Kong. Fruiting period: July to October in Hong Kong.

① 樹幹 TRUNK	② 樹皮 BARK	③ 葉 LEAVES
④ 孢子葉球 POLLEN CONES	⑤ 毬果 SEED CONES	

① 落羽杉具板根。樹幹周邊通常長有膝狀呼吸根（膝根）。樹幹基部腫脹，向上漸尖生長／尖削度大，樹冠圓錐形，枝條橫向開展。

Taxodium distichum (L.) Rich. has buttress roots. Pneumatophores (knees) present or absent neighbouring the trunk. Trunks are Swollen at base, tapering upward and generating a conical tree form, branches extend horizontally. Branchlets distichous.

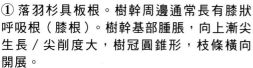

② 樹皮褐色，長條狀剝落。

Bark brown, peeling off in long strips.

③ 葉片互生，當年生小枝上排成 2 列。葉片條狀，扁平，基部扭轉，頂端急尖，呈淡綠色，在冬天轉為暗紅棕色並剝落。

Simple leaves alternate, distichous. Blade linear, base twisted, apex acute, pale green, dark reddish brown and peeling off in winter.

④

⑤

⑤

④ 雌雄同株，單性花。頂生，擁擠，排列成總狀花序或圓錐花序。

Monoecious. Terminal, crowded, arranged in dense racemes or panicles.

⑤ 毬果球形，成熟時由綠色轉為棕黃色或被白色粉末。種子褐色，具翅。

Seed cones globose to ovoid, brownish yellow or white powdery when mature. Seeds brown, winged.

生態 ECOLOGY

植物根系需要氧氣進行有氧呼吸，當氧氣不足時，會出現根部腐爛的情況。那麼，為何落羽杉能在沼澤中存活？物競天擇下，落羽杉進化出「呼吸根」來協助自身在無氧環境中生存。由於呼吸根高度約乎人類膝蓋，故別名為「膝根」。落羽杉的根系被泥土覆蓋，而膝根則裸露在泥面上。落羽杉的木材輕盈且多孔，使氧氣能夠擴散到韌皮部並為根部疏氣。膝根總是暴露在植株根部附近，而膝根的數量取決於植株附近的空間，空間愈大，數量愈多。種植在池塘的落羽杉明顯擁有更多膝根，反觀，栽培在乾燥環境下的落羽杉只有寥寥無幾的膝根。

Plants generally require sufficient oxygen below the ground to avert root rotting. You may wonder, why Bald Cypress can still thrive in swamp forests, the habitats of the tree. For not being ruled out by the nature, Bald Cypress has evolved pneumatophores, a type of modified root that renders the tree excellent acclimatising to anaerobic environments. Due to its comparable height to human knee, it is also called as "knee". Knees are submerged from the main roots of the tree. They are meticulously designed with a rather porous outermost woody layer, allowing oxygen to diffuse easily into phloem and aerate the root system. Knees are always neighbouring the roots while the abundance of it is determined by where the tree grows. Bald Cypress rooting in ponds tends to grow more knees, whereas only a few to no knees are observed in dry soils.

生命力 VITALITY

落羽杉對乾旱、水淹和陰暗環境具高適應能力，但偏好生長於全日照和酸性土壤。

Bald Cypress shows marvellous adaptation to drought, inundated and shady environments. Good planting prefers full sun exposure and acidic soils.

側柏

Chinese Arborvitae, Oriental Arbor-vitae | *Platycladus orientalis* (L.) Franco

相片拍攝地點：香港動植物公園、沙田公園
Tree Location: Hong Kong Zoological and Botanical Gardens, Sha Tin Park

名字由來 MEANINGS OF NAME

屬名 *Platycladus* 意指其屬具扁平的小枝。種加詞 *orientalis* 意指「東方」，暗示其原生於東方國家。

The generic name *Platycladus* refers to its flattened branchlets. The specific epithet *orientalis* means "of the East", alluding to its origin from eastern countries.

本地分佈狀態 DISTRIBUTIONS	外來物種 Exotic species
原產地 ORIGIN	中國中北部和東南部、黑龍江、伯力和北韓。 North-Central and Southeast China, Amur, Khabarovsk and North Korea.
生長習性 GROWING HABIT	常綠喬木。高度可達 20 米。樹冠老時由金字塔形轉為寬卵形。 Evergreen tree. Up to 20 m tall. Crown pyramidal when young, broadly ovoid when old.

1	2	3	4	5	6	7	8	9	10	11	12

花果期月份

花期：本港三月至四月。果期：本港十月。
Flowering period: March to April in Hong Kong. Fruiting period: October in Hong Kong.

① 樹幹 TRUNK	② 樹皮 BARK	③ 葉 LEAVES
④ 孢子葉球 POLLEN CONES	⑤ 毬果 SEED CONES	

① 側柏的樹幹。

Trunk of *Platycladus orientalis* (L.) Franco.

② 樹皮紅棕色至淺灰棕色，長條狀剝落。

Bark reddish brown to light greyish brown, flaking in long strips.

③ 葉片綠色，密集，鱗片狀，4 片排成一平面，具腺點。葉面菱形，頂端鈍尖，與 2 片側葉重疊，葉背中央具明顯凹槽。側葉船形，具脊，頂端稍內彎。

Leaves green, compressed, scaly, 4 arranged in a plane, covered with glandular dots. Facial leaves rhombus, apex bluntly pointed, overlapped by 2 lateral leaves, abaxially with an eminent groove at central. Lateral leaves boat-shaped, ridged, apex slightly incurved.

④ 卵形，呈黃綠色。雌雄同株。

Ovoid, yellowish green. Monecious.

⑤ 球形，成熟時由白綠色轉為紅棕色並裂開。種子為卵形，淡褐色，無翅。

Globose, turning whitish green to reddish brown and ruptured at maturity. Seeds ovate, light brown, wingless.

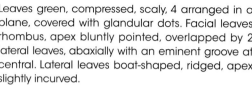

側柏是一種多用途的觀賞樹種，常被種植於公園和花園內。其樹幹筆直，頂部漸尖，故常被單獨種植以呈現其優美的樹形。鑑於其樹枝柔軟且可彎曲，故常被扭曲成不同造型以供觀賞。

側柏的木材柔軟且堅固耐用，故可加工成家具、船舶和木雕。

Chinese Arborvitae is a versatile ornamental tree serving multiple functions in parks and gardens. The tree is always planted in solitary for its straight stem and tapering head. On account of its soft and malleable branches, the tree is always twisted into different decorative shapes.

The wood is soft, durable and strongly resistant. Hence, it can be processed into furniture, ships and wooden carvings.

生態 ECOLOGY

側柏通常可以存活幾個世紀，其長壽一直備受大眾讚賞。最古老的側柏位於中國陝西省，其樹齡約有 4000 年。全靠其松柏之壽，才能讓我們更了解人類歷史和重要的進化事件。

側柏被視為絕佳的觀賞樹種並廣泛種植於公園和花園。儘管側柏算是世上的優勢種，但不幸地，野生側柏數量正在減少。根據近年的分子研究顯示，目前中國古側柏的數量已分裂成幾個小規模的種群。分化的原因可歸究於棲息地於短時間內大量喪失，原本密集的棲息地變得支離破碎。一旦側柏的數量減少，便會助長了近親繁殖和遺傳漂變，這會淘汰掉基因多樣性並增加其滅絕的風險。最終，側柏被針葉樹專家組評為「近危」。然而，由於我們對非遠古及野生側柏的認知匱乏，故其保育狀態仍備受爭議。為保護野生側柏，我們迫切需要更多的研究並制定保育行動。

Chinese Arborvitae is always appreciated for its longevity that can generally survive for centuries. The oldest Chinese Arborvitae has already reached approximately 4000 years in the Shaanxi Province of China. By virtue of its rather longer lifespan, the ancient population of Chinese Arborvitae has opened avenues for our glimpsing of human history and early evolutionary events.

Chinese Arborvitae is an excellent ornamental tree widely cultivated in parks and gardens. Despite its ubiquity in urban areas, the wild population is unfortunately shrinking. Recent molecular studies showed that the current population of ancient Chinese Arborvitae in China has been fragmented into a few groups with smaller population size. Probably, intensive habitat loss and fragmentation could attribute to this divergent event. Once a population size is getting reduced, it always encourages inbreeding and genetic drift, the events that can eliminate genetic diversity and increase the probability of extinction. As a result, Chinese Arborvitae is rated as "near threatened" by the Conifer Specialist Group. However, the conservation status is still in dispute due to a scarce understanding of its non-ancient population. More studies and conservation actions are urgently needed to preserve the tree.

竹柏

Nagi, Japanese Podocarpus │ *Nageia nagi* (Thunb.) Kuntze

相片拍攝地點：香港公園、荃灣公園、沙田公園
Tree Location: Hong Kong Park, Tsuen Wan Park, Sha Tin Park

名字由來 MEANINGS OF NAME

Nagi（ナギ）在日文中蘊含平靜的意思。由於竹柏的平行葉脈不明顯，與竹子十分相似，加上其樹皮材質與柏相像，故有竹柏此名。

The common name "Nagi" is derived from the Japanese word, referring to "calm". The Chinese name「竹柏」describes its leaves sharing the parallel veins as bamboo (竹) leaves while the bark texture is similar to juniper (柏).

本地分佈狀態 DISTRIBUTIONS	外來物種 Exotic species
原產地 ORIGIN	日本，亦廣泛分佈於中國中部、西南部、東南部、南部等省份。 Japan, also widely distributed in Central, Southwestern, Southeastern and Southern China.
生長習性 GROWING HABIT	常綠喬木。高度可達 20 米。 Evergreen tree. Up to 20m tall.

花果期 月份	1	2	3	4	5	6	7	8	9	10	11	12

花期：本港三月至五月。果期：本港八月至十一月。
Flowering period: March to May in Hong Kong. Fruiting period: August to November in Hong Kong.

竹柏的木材、葉片和種子具多種用途。竹柏木質細膩耐用，可用於製作家具、建築用料和容器。此外，木材萃取物可用於製作織物染料。其葉片具藥用價值，常用於止血和治療骨折。竹柏的種子則含有豐富油脂，可食用。

由於竹柏具堅實的樹形和闊大樹冠，故被廣泛種植作行道樹和觀賞樹，亦是盆景栽培的常見品種。

Nagi has been exploited for multiple applications with its wood, leaves and seeds. Its wood is fine and durable, and thus always used for making furniture, containers and other construction; the wood extracts can be used for fabric dyes. Its leaves are the primary sources of a traditional Chinese medicine for treating hemostasis and bone fractures. The seeds contain rich oil content and are edible.

By virtue of its handsome tree form with dense foliage, decent crown and prominent stem, Nagi is always yearned for shading and ornamental purposes. It is also a prevalent species in bonsai cultivation.

辨認特徵 TRAITS FOR IDENTIFICATION

① 樹幹 TRUNK	② 樹皮 BARK	③ 葉 LEAVES
④ 孢子葉球 POLLEN CONES	⑤ 毬果 SEED CONES	

① 或具板根。小枝直立、開展。

Sometimes with buttress root. Branchlets erect and spreading.

② 樹皮紅棕色至深紫紅色，呈小片厚片狀剝落。

Bark reddish brown to dark purplish red colour, peeling in small, thick flakes.

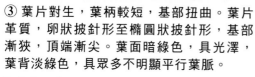

③ 葉片對生，葉柄較短，基部扭曲。葉片革質，卵狀披針形至橢圓狀披針形，基部漸狹，頂端漸尖。葉面暗綠色，具光澤，葉背淡綠色，具眾多不明顯平行葉脈。

Petioles short and strongly twisted at base. Leaves opposite. Blade leathery, ovate-lanceolate to elliptic-lanceolate, base attenuate, apex acuminate, dark green and glossy adaxially, pale green abaxially, with many indistinct parallel veins.

④ 雌雄異株，毬果腋生。雄毬果圓柱形，卵球形，短而粗。雌性毬果單生，很少成對，球狀。

Dioecious. Cones axillary. Male cones cylindric, ovoid-cylindric, short and thick. Females cones solitary, rarely paired, globose.

⑤ 種子圓球形至梨形，基部尖，頂端圓形，被白色粉末，成熟時從由綠色轉為暗紫色。

Globose to pyriform, base pointed, apex rounded, covered with whitish powder, turning green to dark purple when ripe.

生命力 VITALITY

此物種對陰暗環境具有良好的耐受性，同時偏好疏水、透氣和濕潤的土壤。

The tree is rather shade-tolerant and prefers well-drained, aerated and moist soils.

含笑

Banana Shrub | *Michelia figo* (Lour.) Spreng.

相片拍攝地點：香港動植物公園、遮打花園
Tree Location: Hong Kong Zoological and Botanical Gardens, Chater Garden

名字由來 MEANINGS OF NAME

含笑的花香氣撲鼻，聞起來有陣陣香蕉甜膩的氣味，因此被稱為「香蕉花」。有趣的是，花朵往往會開成兩半，就像給人一種靦腆含蓄的微笑，故亦有「含笑」此美稱。

Its flowers are fragrant and smelled like a banana, hence named as "Banana Shrub". Interestingly, instead of blossoming completely, its flowers tend to be not spreading, just like giving an implicit smile which can be directly translated into「含笑」in Chinese.

應用 APPLICATION

含笑的葉片茂密，經常被修剪成樹籬或灌木叢，用作天然籬笆或裝飾用途。含笑的花香氣撲鼻，能提煉成精油，發揮抗菌作用，也可加工成花茶。其花蕾在中醫學上能治理月經不調。它的葉片能用作外敷，治理跌傷。

Banana Shrub has dense foliage and is always shaped into hedges or bushes for a great deal of functional or decorative purposes. Its flowers are fragrant and the extracts are antimicrobial. The flowers are also esculent and can be processed into flowering tea. The flower buds are the primary source of traditional Chinese medicines for treating abnormal menstrual periods. The leaves are externally used for treating injuries from falls.

本地分佈狀態 DISTRIBUTIONS	外來物種 Exotic species
原產地 ORIGIN	廣東、福建等華南省份。 The provinces of Southeast China such as Guangdong and Fujian.
生長習性 GROWING HABIT	常綠灌木。高度可達 3 米。 Evergreen shrub. Up to 3 m tall.

1	2	3	4	5	6	7	8	9	10	11	12

花果期月份

花期：本港三月至五月。果期：本港七月至八月。
Flowering period: March to May in Hong Kong. Fruiting period: July to August in Hong Kong.

① 樹幹 TRUNK	② 樹皮 BARK	③ 葉 LEAVES
④ 花 FLOWERS	⑤ 果 FRUITS	

① 含笑的樹幹。

Trunk of *Michelia figo* (Lour.) Spreng.

② 樹皮暗灰色，稍粗糙，分枝眾多。

Bark dark grey, slightly rough, multi-branched.

③ 葉片互生，狹橢圓形或倒卵狀橢圓形，頂端鈍銳形，基部楔形或寬楔形，葉面光滑無毛，葉背中脈具棕色平伏毛。嫩枝、芽、葉柄和花梗均被黃棕色短柔毛。托葉痕長達葉柄頂端。

Leaves alternate. Blade narrowly elliptic or obovate-elliptic, apex obtusely acute, base cuneate or broadly cuneate, adaxially glossy and glabrous, abaxial midvein covered with brown appressed hairs. Young twigs, buds, petioles and peduncles are covered by yellowish brown pubescence. Long stipular scar almost reaching the apex of petiole.

④ 花被片 6 片，呈淡黃色，基部和葉緣有時呈紅色或紫色。雌蕊柄顯眼，呈淡黃色。花香甜美。

Tepals 6, pale yellow but margin sometimes red to purple. Gynophore petiole conspicuous, pale yellow, sweetly fragrant.

⑤ 蓇葖果卵圓形或球形，先端具小喙。

Follicles ovoid or globose, apex with a small beak.

含笑是其中一種統帥青鳳蝶和木蘭青鳳蝶的幼蟲食用的植物。

The leaves of Banana Shrub serve as one of the larval food plants of *Graphium agamemnon* (Tailed Jay) and *Graphium doson* (Common Jay).

此物種對陰暗環境具有良好的耐受性，但偏好全日照和溫暖的生長環境。

Banana Shrub shows excellent tolerance to shade, but good planting prefers full sunlight and warm environments.

二喬木蘭 又稱：紫玉蘭

Saucer Magnolia | *Yulania × soulangeana* (Soul.-Bod.) D. L. Fu

相片拍攝地點：香港動植物公園、城門谷公園
Tree Location: Hong Kong Zoological and Botanical Gardens, Shing Mun Valley Park

名字由來 MEANINGS OF NAME

為紀念法國植物學家 —— 艾蒂安 · 蘇朗熱 · 博丹（1774-1846）曾將玉蘭與紫玉蘭雜交，屬名特此取其名，並為二喬木蘭冠名為 *soulangeana*。

The specific epithet *soulangeana* commemorates Etienne Soulange-Bodin (1774-1846), a French botanist who astonishingly hybridized the tree from *Magnolia denudata* (Yulan Magnolia) and *Magnolia liliiflora* (Purple Magnolia).

本地分佈狀態 DISTRIBUTIONS	外來物種 Exotic species
原產地 ORIGIN	中國中南部。 South-Central China.
生長習性 GROWING HABIT	落葉喬木。高度可達 10 米。 Deciduous tree. Up to 10m tall.

花果期 月份	1	2	3	4	5	6	7	8	9	10	11	12

花期：本港二月至三月。果期：本港九月至十月。
Flowering period: Februaruy to March in Hong Kong. Fruiting period: September to October in Hong Kong.

　　二喬木蘭的花朵妊紫嫣紅，故被視為觀賞高的樹種。立春之時，二喬木蘭艷麗的化朵盛放，吸引途人佇足鑑賞。令人賞心悅目的花和芳香的氣味隱若地提醒着大家生機勃勃的春天要來臨了。然而，開花時間可能因品種而異。根據我們的觀察，部分二喬木蘭在秋季或更早時段開花。

　　Saucer Magnolia is highly valued as an ornamental tree by virtue of its spectacular blossoms. In the beginning of spring, the tree blossoms and the flamboyant floral display immediately to draw visitors' attention. Its delightful perianth colour and aromatic smell seems mildly remind people the arrival of spring, the season of resilience. However, the flowering time could be varied among cultivars. According to our observations, some Saucer Magnolia blossom in autumn or even early.

辨認特徵 TRAITS FOR IDENTIFICATION

① 樹幹 TRUNK	② 樹皮 BARK	③ 葉 LEAVES
④ 花 FLOWERS	⑤ 果 FRUITS	

① 二喬木蘭的樹幹。

Trunk of *Yulania × soulangeana* (Soul.-Bod.) D. L. Fu

② 樹皮淡灰色，光滑，具皮孔，小枝無毛，具葉痕。

Bark light grey, smooth, lenticellate. Branchlets glabrous, with leaf scars.

③ 葉片紙質，葉背被柔毛，倒卵形，頂端短銳形，基部楔形，葉緣全緣，漸變窄，側脈 7-9 條。葉柄被短柔毛，托葉痕約為葉柄長度的 3 分之 1。

Leaves alternate, papery, abaxially pubescent, obovate, apex short acute, base cuneate, entire, narrowed gradually, lateral veins 7-9. Petiole pubescent, with stipular scar around 1/3 as long as petiole.

④ 完全花。花蕾卵形，呈黃綠色，密被柔毛。花朵先於葉片出現。花被片呈淡紅色至暗紅色，外輪 3 片花被片較短，約為內輪花被片長度的 3 分之 2。雄蕊眾多，側面裂開，雌蕊圓柱形，心皮離生，眾多。

Hermaphroditic. Flower buds ovoid, yellowish green, covered with dense hairs. Flowers appearing before leaves. Tepals pale red to dark red, outer 3 tepals always 2/3 as long as inner ones. Stamens many, laterally dehiscent, gynoecium cylindrical, apocarpous, many.

⑤ 聚合蓇葖果，卵形或倒卵形，成熟時轉為黑色。種子深褐色，側面側扁。

Aggregate fruit, follicles, ovoid or obovoid, black when ripe. Seed dark brown, obovate, compressed laterally.

植物趣聞 ANECDOTE ON PLANTS

木蘭科 Magnoliaceae：

從植物分類學的角度而言，木蘭科是由許多基部開花植物組成。其花朵原始的結構與菊科較高等的花朵結構截然不同。以下是木蘭科的原始特徵，包括：（1）花瓣與萼片融合且不明顯，此結構名為花被片；（2）雄蕊呈螺旋狀簇生在較大的花托上，雌蕊位於其上具許多與別不同的心皮；（3）大型單生花，芬香迷人；（4）花朵被苞片包裹。

花的形態是鑑定木蘭科植物的關鍵。例如，較大的白色花朵通常是荷花玉蘭或玉蘭。二喬木蘭與紫玉蘭的花朵皆姹紫嫣紅。要區分兩者，可從花被片及莖入手。二喬木蘭的外輪花被片較短，約為內輪花被片的 3 分之 2。而紫玉蘭的外輪花被片更短，且呈萼片狀。此外，二喬木蘭多數具明顯的莖，像樹一般；紫玉蘭多數像灌木般矮小，香港很少種植，大部分標記為紫玉蘭的應該是二喬木蘭。

From the perspective of plant taxonomy, Magnoliaceae is a family composed of many basal flowering plants whose flower structures are rather primitive and differed from those in advanced families (e.g. Asteraceae). The primitive characteristics are: 1) their petals and sepals are fused and indistinct (termed as tepals); 2) androecium are spirally clustered at an enlarged receptacle while gynoecium is located at above with many unfused carpels; 3) solitary flowers are large, fragrant and glamorous; 4) the flowers are enclosed in bracts.

Floral morphology is the key to identify members of Magnoliaceae. For example, sizable flowers in white always refer to *Magnolia grandiflora* (Southern Magnolia) or Yulan Magnolia. The flowers of Saucer Magnolia and Purple Magnolia are both majestic and purple. To distinguish the trees, the outer tepals of Saucer Magnolia are 2/3 of the inner ones, while those of Purple Magnolia are much shorter and sepal-like. Moreover, Saucer Magnolia is always tree-like with a prominent stem, whereas Purple Magnolia is shrubby. In Hong Kong, Purple Magnolia is rarely planted and most of them marked as Purple Magnolia should be instead Saucer Magnolia.

黃樟 又稱：黃樟木

Yellow Cinnamomum, Yellow Camphor-tree | *Cinnamomum parthenoxylon* (Jack) Meisn.

相片拍攝地點：香港仔樹木研習徑、寶雲道健身徑、城門水塘、獅子會自然教育中心
Tree Location: Aberdeen Tree Walk, Bowen Road Fitness Trail, Shing Mun Reservoir, Lions Nature Education Centre

應用 APPLICATION

　　黃樟的木材和果實具濃烈香樟味能驅蟲。黃樟木和葉能加工成樟腦油，用作天然家居驅蟲劑。由於黃樟木結實，擁有抗蟲抗菌的特質，故被廣泛用作造船、製作家具和藝術品。黃樟種子可加工作肥皂。此外，黃樟全株均可入藥，用作醫治類風濕性關節炎及痢疾。鑑於其樹冠闊大，大樹可乘涼，故目前此樹種在街道上被廣泛種植。

本地分佈狀態 DISTRIBUTIONS	原生物種 Native species
原產地 ORIGIN	遍佈於中國南部及西南，同時廣泛分佈於印度、馬來西亞、印尼、巴基斯坦和澳洲南部。 It is widely distributed in Southern and Southwestern China, as well as in India, Malaysia, Indonesia, Pakistan and Southern Australia.
生長習性 GROWING HABIT	常綠喬木。高度可達 20 米。 Evergreen tree. Up to 20m tall.

1	2	3	4	5	6	7	8	9	10	11	12

花果期月份

花期：本港三月至五月。果期：本港四月至十月。
Flowering period: March to May in Hong Kong. Fruiting period: April to October in Hong Kong.

The wood and fruit of Yellow Cinnamomum have a strong aroma of camphor to repel insects. The essential oils extracted from its wood and leaves, known as "camphora oil", are competent to expel household pests. Since the wood is sturdy and highly resistant to insects and fungi, it is yearned for making ships, furniture and artworks. The seeds can also be processed into soaps. Moreover, the tree serves versatile medicinal functions and is commonly processed into traditional Chinese medicines to treat rheumatoid arthritis and dysentery. Regarding its straight trunk and large canopy, Yellow Cinnamomum is a superb street tree species for providing shade effect.

辨認特徵 TRAITS FOR IDENTIFICATION

① 樹幹 TRUNK	② 樹皮 BARK	③ 葉 LEAVES
④ 花 FLOWERS	⑤ 果 FRUITS	

① 黃樟的樹幹。

Trunk of *Cinnamomum parthenoxylon* (Jack) Meisn.

② 樹皮呈綠棕色，具深縱裂，片狀剝落。整株具強烈的樟腦氣味。

Bark greenish brown, with deep longitudinal fissures, peeling off in small flakes. The entire plant strongly camphor-scented.

③ 葉片互生，橢圓狀卵形或長橢圓狀卵形。頂端常急尖或短漸尖。基部楔形或闊楔形，葉緣全緣，革質，兩面無毛，葉背淺綠色。羽狀脈，葉面脈腋具不明顯突起，壓碎時釋出樟腦味。

Simple leaves alternate. Blade leathery, glabrous, adaxially dark green and lustrous, abaxially pale green, elliptic-ovate or long elliptic-ovate, apex always acute or shortly acuminate, base cuneate to broadly cuneate, entire, pinnately veined, inconspicuous bulges at axils of lateral veins. Smell of camphor when crushed.

④ 圓錐花序或聚傘花序。花朵細小，呈綠黃色。

Panicles or cymes axillary on branchlets. Flowers small, greenish yellow.

⑤ 果球形，成熟時轉為黑色。

Drupes globose, turning black at maturity.

植物趣聞 ANECDOTE ON PLANTS

黃樟和樟 Yellow Cinnamomum and *Cinnamomum camphora* (Camphor Tree)：

黃樟和樟皆為樟屬，兩者特質相似，難以分辨。雖被稱作黃樟，但它不具明顯黃色。可從葉脈入手分辨兩者。除了樟屬的共同特質：具離基三出脈或三出脈外，黃樟的羽狀脈和互生葉序是另一種辨認特質。值得關注的是，樟樹毫無疑問是香港最普遍的城市樹木之一，而黃樟是罕見種植和相當局限於其野生棲息地。然而，由於其密集的葉片和優良的耐病原體能力，黃樟仍然是一個潛在的綠化組成樹種。

Both species are the members of *Cinnamomum* and share imperceptible appearances. The most critical trait to differentiate is that the leaves of Camphor Tree are triplinerved, while those of Yellow Cinnamomum are pinnately veined and arranged alternately on leaves. Notably, Camphor Tree is beyond all dispute one of the most pervasive urban trees in Hong Kong, whereas Yellow Cinnamomum is uncommonly planted and rather restricted to its wild habitats. However, the tree is still a potential greening component by virtue of its dense foliage and excellent tolerance to pathogens.

浙江潤楠 又稱：長序潤楠

Chekiang Machilus, Zhejiang Machilus | *Machilus chekiangensis* S. K. Lee

相片拍攝地點：川龍、大埔滘
Tree Location: Chuen Lung, Tai Po Kau

應用 APPLICATION

　　浙江潤楠的枝葉可入藥，能消腫、化痰、活血、止痛。它也用於治療支氣管炎和燙傷。

Its branches and leaves are valued for versatile medicinal functions; they are usually used for dispersing swelling, relieving phlegm, blood and pain, and curing bronchitis and scalds.

本地分佈狀態 DISTRIBUTIONS	原生物種 Native species
原產地 ORIGIN	浙江、福建。 Zhejiang, Fujian.
生長習性 GROWING HABIT	常綠喬木。高度可達 10 米。 Evergreen tree. Up to 10 m tall.

花果期 月份

1	2	3	4	5	6	7	8	9	10	11	12

花期：本港二月。果期：本港四月至五月。
Flowering period: February in Hong Kong. Fruiting period: April to May in Hong Kong.

① 樹幹 TRUNK	② 樹皮 BARK	③ 葉 LEAVES
④ 花 FLOWERS	⑤ 果 FRUITS	

① 浙江潤楠的樹幹。幼枝基部具芽鱗痕。

Trunk of *Machilus chekiangensis* S. K. Lee. Young branches with notable bud scale scars at the base.

② 樹皮褐色，唇形皮孔散生。

Bark brown, with scattered labiate lenticels.

③ 葉片互生，常聚生於小枝頂端。葉片倒披針形，基部漸狹，頂端常驟尖，幼時被微柔毛，後變無毛，成熟時革質或薄革質，葉背中脈明顯凸起。大芽鱗是浙江潤楠的一種特徵。

Simple leaves alternate, commonly aggregated at the apex of branchlets. Leaf blade oblanceolate, base tapering, apex often cuspidate, puberulent when young, becoming glabrous later, leathery or thinly leathery when mature, midrib of abaxial leaf apparently raised. Large bud scale is a characteristic of Chekiang Machilus.

④ 圓錐花序，花序生於當年生的樹枝基部。花被片 6 片，呈黃綠色，被微柔毛。

Panicles at the base of current year's branchlets. Tepals 6, yellow green, puberulent.

⑤ 核果，球狀，成熟時轉為黑色。

Drupes globose, fleshy, black at maturity.

生態 ECOLOGY

　　浙江潤楠屬早期演替品種，生長迅速，對光照要求高。它和其他潤楠屬，例如短序潤楠和刨花潤楠已取代次生林中的不耐蔭品種，並在香港次生林樹冠中佔有相當大的比例。

Chekiang Machilus is known as an early successional tree which is fast growing and light demanding. Including Chekiang Machilus, *Machilus* spp. such as *Machilus breviflora* (Short-flowered Machilus) and *Machilus pauhoi* (Many-nerved Machilus) have occupied a discernible proportion of tree canopy of the forests and ruled out the species which grow rampantly in grasslands and are less acclimatised to shady environments.

生命力 VITALITY

　　浙江潤楠對陰暗及貧瘠土壤具高耐受性。

Chekiang Machilus shows excellent tolerance to shade and barren soils.

菠蘿蜜 又稱：樹菠蘿、木菠蘿

Jackfruit | *Artocarpus heterophyllus* Lam.

相片拍攝地點：鰂魚涌公園、香港公園
Tree Location: Quarry Bay Park, Hong Kong Park

名字由來 MEANINGS OF NAME

聽到「菠蘿蜜」這大名，相信大家會聯想到家傳戶曉的大作《西遊記》—— 唐僧所背誦佛經，Prajñāpāramitāhṛdaya 是其梵文名稱，《般若波羅蜜多心經》則為其中文名稱，有「智慧圓滿之心」的意思，其中梵文 paramitā 也可翻譯為「到達彼岸」或「步向圓滿」。不過，這聽起來與樹木毫無瓜葛。反而，《本草綱目》記載：「波羅蜜，梵語也。因此果味甘，故借名之。」，說明「菠蘿蜜」這名稱可能源自其甜味，而不是佛經中的意思。

雖然此品種的名稱似乎不是源自佛教，但這樹種實際上離印度 —— 佛教起源地很近。在印度，菠蘿蜜最初種植在一座佛教寺廟中，其葉片和木材具多種用途。例如收集回來的葉片經常用於印度教寺廟禮拜中；從樹上提取的木材用於建造寺廟。無可否認，這樹種與佛教之間，可能存在着千絲萬縷的「因緣」。

本地分佈狀態 DISTRIBUTIONS	外來物種 Exotic species
原產地 ORIGIN	印度。 India.
生長習性 GROWING HABIT	常綠喬木。高度可達 20 米。 Evergreen tree. Up to 20m tall.

1	2	3	4	5	6	7	8	9	10	11	12

花果期月份

花期：本港三月至八月。果期：本港六月至十一月。
Flowering period: March to August in Hong Kong. Fruiting period: June to November in Hong Kong.

Prajñāpāramitāhṛdaya (般若波羅蜜多心經 in Chinese) is a Sanskrit title of renowned Buddhist text with which we always associate「菠蘿蜜」. It can be sense-for-sense translated into "the heart of the perfection of wisdom". The Sanskrit word, *paramita*, can be also understood as "reaching the opposite shore" or "perfection". However, any of the above elaborations sound irrelevant to the tree. The sweet taste is rather pertinent. As what the *Compendium of Materia Medica* (*Bencao Gangmu*) depicted, "*pāramitāā*, Sanskrit. Its fruit is sweet, hence named in this way". Here shows that「菠蘿蜜」is more congruent to the sweetness of the fruit, rather than its meaning in the Buddhist text.

Although the tree is less pertinent to Buddhism, it is actually close to the religion in daily life. In India, Jackfruit trees have been early planted in Buddhist temples and served multiple uses with its leaves and woods. For example, the leaves are collected for worshiping in Hindu temples. Moreover, the wood extracted from the tree is always used for temple construction. We have to say, there could be a deep and mysterious *Nidāna* between the tree and Buddhism.

辨認特徵 TRAITS FOR IDENTIFICATION

① 樹幹 TRUNK	② 樹皮 BARK	③ 葉 LEAVES
④ 花 FLOWERS	⑤ 果 FRUITS	

① 菠蘿蜜的樹幹。年老時具板根。

Trunk of *Artocarpus heterophyllus* Lam. Buttressed when old.

② 樹皮厚，呈黑棕色，全株具乳膠。

Bark thick, blackish brown, latex throughout.

③ 葉片輪生，厚革質，光滑，無毛，葉面呈暗綠色，葉背呈淡綠色，稍粗糙，橢圓形或倒卵形，基部楔形，頂端鈍形或漸尖。

葉緣全緣，幼樹或萌發枝常分裂。側脈 6-8 對，中脈葉背明顯凸起。托葉卵形，環狀抱莖，早落，脫落後形成托葉環痕。

Leaves spirally arranged, blade thickly leathery, smooth, glabrous dark green adaxially, slightly rough and pale green abaxially, elliptic to obovate, base cuneate, apex obtuse to acuminate, margin entire, young trees or saplings often divided. Lateral veins 6-8 pairs, midvein obviously prominent abaxially. Stipules ovate, caducous.

④

④

⑤

④ 單性花，雌雄同株。雄性花序圓柱狀或棒狀圓柱形，腋生於小枝，成熟時轉為暗綠色。雌性花序呈橢圓形或圓形，具花萼管，集中於樹幹或主枝，初時密閉，形似佛焰苞。

Monoecious. Male inflorescences cylindrical or club shape, axillary on branchlets, darker green when mature. Female inflorescences elliptic or rounded. Calyx tube, concentrated on trunk or main branch, initially closed, resembling spathe.

⑤ 大型聚花果，呈黃棕色，橢圓狀至球狀或不規則狀，具六角形瘤狀凸體及粗絨毛。成熟時由黃綠色轉為黃棕色。

Multiple fruit (syncarps) large, yellow-brown colour, ellipsoid to globose or irregular, composed of stiff hexagonal tubercles. Turning yellowish green to yellowish brown when mature.

應用 APPLICATION

菠蘿蜜的木材、葉片、果實和種子具有多種用途。菠蘿蜜的木材廣泛應用在製作絲綢染料、棉袍和家具。此外，其葉片是治療發燒和其他皮膚病的重要藥材。它還可以大幅降低患上高血壓、心臟病和中風的風險。果實成熟後能成為美味佳餚，亦可加工成果醬、果凍、麵條和糖果。種子多汁，通常經煮沸或烘烤後加工成食用堅果。種子含大量澱粉質，能舒緩膽汁分泌過多的情況。

Jackfruit serves versatile functions with its wood, leaves, fruits and seeds. The wood is widely applied for making dye silk, cotton robes, and furniture. The leaves are an essential ingredient to treat fever and other skin diseases. It can also reduce the risk of high blood pressure, heart diseases and strokes. The mature fruits are considered as a great delicacy that can be eaten flesh but often processed into jam, jelly, pasta and candies. The seeds are esculent and can be prepared into table nuts after boiling or roasting. Seeds contain a lot of starch; can ease the excessive secretion of bile.

構樹 又稱：鹿仔樹

Paper Mulberry | *Broussonetia papyrifera* (L.) L'Hér. ex Vent.

相片拍攝地點：香港中文大學、黃泥頭
Tree Location: The Chinese University of Hong Kong, Wong Nai Tau

名字由來 MEANINGS OF NAME

　　古時候，人們會以構樹的嫩葉餵飼鹿、牛和羊，故有「鹿仔樹」這個中文名稱。由於構樹的果實與桑樹的果實相似，而構樹的葉片粗糙而呈紙質，故又稱「紙桑」。

　　屬名 *Broussonetia* 是為了紀念法國自然學家布魯桑尼特（1761-1807），而種加詞 *papyrifera* 描述此品種的葉片及樹皮可用於製造紙張。

本地分佈狀態 DISTRIBUTIONS	原生物種 Native species
原產地 ORIGIN	中國華東、華南和西南省份。同時原產於印度和東南亞國家，如老撾、越南等。 East, Southern and Southwestern China. Also originated in India and Southeast Asia, e.g. Laos and Vietnam.
生長習性 GROWING HABIT	常綠喬木。高度可達 20 米。 Evergreen tree. Up to 20 m tall.

花果期 月份	1	2	3	4	5	6	7	8	9	10	11	12

花期：本港三月至五月。果期：本港四月至八月。
Flowering period: March to May in Hong Kong. Fruiting period: April to August in Hong Kong.

Since people in the past loved feeding deer, cattle and sheep with Paper Mulberry's young leaves, the tree is also named as 「鹿仔樹」. Its leaves are thick papery but with the fruits that are outwardly similar to those of *Morus alba* (White Mulberry), hence named as "Paper Mulberry".

The generic name *Broussonetia* is named after Pierre Auguste Marie Broussonet (1761-1807), who was an illustrious French naturalist from Montpellier. The specific epithet *papyrifera* describes the use of the bark for paper making.

應用 APPLICATION

　　構樹因其豐富的經濟價值而被廣泛種植。首先，其樹皮可用於製造紙張。在日本，樹皮的內皮是「和紙」的主要材料。中國棉紙主要是由結香樹皮、雁皮樹皮和構樹樹皮構成。其樹皮纖維較硬，可加工成繩索。木材質地輕巧，故常用作製作家具。由於其果實含有豐富的皂甙和維他命 B，故被用作增強記憶力和緩解阿茲海默症的中藥材料。

Paper Mulberry is widely planted for prolific economic values. First, its bark is useful for paper making. In Japan, the inner bark is the chief material of washi. The Chinese cotton paper is made of the barks of *Edgeworthia chrysantha* (Oriental paperbush), *Wikstroemia sikokiana* and Paper Mulberry. The bark fibre is stiff and can be processed into rope. The wood is light with appealing texture and highly demanded for making furniture. Other than paper making, the fruits are wholesome and contains profuse saponins and B vitamins, which are effective for enhancing memory and attenuating dementia.

辨認特徵 TRAITS FOR IDENTIFICATION

① 樹幹 TRUNK	② 樹皮 BARK	③ 葉 LEAVES
④ 花 FLOWERS	⑤ 果 FRUITS	

① 構樹的樹幹。

Trunk of *Broussonetia papyrifera* (L.) L'Hér. ex Vent.

② 樹皮灰棕色，光滑並具縱裂紋。嫩枝密被絨毛。

Bark taupe, smooth, with longitudinal cracks. Branchlets densely pubescent.

③ 托葉卵形，葉柄被柔毛。單葉互生。葉形多變，寬卵形或長橢圓狀至淺裂，厚紙質，基部心形及不對稱，頂端漸尖，葉緣具細鋸齒或 3-5 裂。

Petioles hairy. Stipules ovate. Simple leaves alternate. Blade thick chartaceous, largely variable, broad-ovate or oblong-ovate, base cordate and unequal, apex acuminate, margin serrulate or 3–5 lobed.

④ 雌雄異株，單性花。雄性花序為葇荑花序，圓柱狀。雌性花序為頭狀花序，球狀，呈綠色，花柱凸出。

Dioecious. Male catkins cylindrical. Female inflorescences head-like, globose, greenish, protruded styles.

⑤ 聚花果，球狀，肉質，瘦果眾多，成熟時轉為紅色。

Syncarps, globose, composed of many achenes, fleshy, turning vermillion at maturity.

生態 ECOLOGY

我們會用「無處不在」來形容構樹，像雜草一樣肆意生長，您可以不費吹灰之力地在道路、灌木叢或森林中找到它（大多為幼年大小）。與其他雜草品種一樣，構樹擁有以下生存策略。首先，此品種生長速度極快。通常在 2 年內長到 3-4 米高。此外，構樹可以通過風傳播花粉，促進遠距離傳播。種子肉質，受許多鳥類和昆蟲的喜愛，這些動物亦促進遠距離傳播。加上對惡劣環境，如乾旱、貧瘠土壤和氣候（從溫帶氣候到熱帶氣候）具高耐受性，故具高適應力的構樹，在許多國家被視為入侵物種。

"Ubiquitous" is how we describe Paper Mulberry. The tree just grows rampantly so you can find it effortlessly along roads, thickets or forests. Comparable with other weedy species, Paper Mulberry has armed the survival strategies as follows. First, it grows drastically fast. Usually, it can attain a mature size (3-4 m) within 2 years. Second, its pollens are dispersed by wind which guarantees a long-distance mating. Third, the seeds are fleshy and beloved by many birds and insects which are motivated vehicles to drive the seeds to everywhere. Coupled with its dramatic resilience to versatile environments (e.g. drought and barren soils) and climates (ranging from temperate to tropical climates), Paper Mulberry is rampantly naturalized and considered as an invasive species in many countries.

花葉垂葉榕 又稱：斑葉垂榕
Ficus benjamina L. 'Variegata'

相片拍攝地點：鰂魚涌公園、荃灣公園、柴灣公園、荔枝角公園
Tree Location: Quarry Bay Park, Tsuen Wan Park, Chai Wan Park, Lai Chi Kok Park

名字由來 MEANINGS OF NAME

栽培種加詞 Variegata 用單引號表示，以表明其葉片顏色遺傳自野生垂葉榕。

The cultivar name 'Variegata', giving in single quotation marks, depicts its variegated leaves which are a cultivated characteristic distinct from wild *F. benjamina*.

應用 APPLICATION

花葉垂葉榕的葉片圖案獨特，故比起垂葉榕更常被種植於花園和公園作觀賞用途。它能提供許多不同風貌的景色，大大提高其觀賞價值。它既可單獨種植，亦可與其他植物一併種植。此外，它的可塑性高，常用於修剪成特殊形狀或樹籬。

Due to its unique foliage pattern, the tree is more often planted in gardens and parks for ornamental than Ficus benjamina L. It can provide many scenic styles, thus greatly improving its ornamental value. It can be planted in solitary or mixed in different greening configurations. In addition, it is highly malleable and is often used for pruning into special shapes or hedges.

本地分佈狀態 DISTRIBUTIONS	外來物種 Exotic species
原產地 ORIGIN	野生垂葉榕原生於中國南部及西南部、南亞和澳洲。 Its wild type (*Ficus benjamina*) is native to South and Southwest China, South Asia and Australia.
生長習性 GROWING HABIT	常綠喬木。高度可達 20 米。 Evergreen tree. Up to 20 m tall.

1	2	3	4	5	6	7	8	9	10	11	12	花果期 月份

花期：甚少開花。果期：甚少結果。
Flowering period: Rarely flowering. Fruiting period: Rarely fruiting.

① 樹幹 TRUNK　　② 樹皮 BARK　　③ 葉 LEAVES

① 具板根，氣根從老枝垂下，到達地面後轉為木質。

Buttressed roots, aerial roots pendulous from old branches, lignified after reaching ground.

② 樹皮灰色，光滑，側枝延長，下垂，無毛。

Bark grey, smooth, lateral branchlets elongated, drooping, glabrous.

③ 單葉互生。葉片呈薄革質，葉面具光澤，綠色減少成鑲嵌圖案，葉片卵形至卵狀橢圓形，頂端漸尖、下垂，葉緣全緣和波狀，側脈眾多。

Simple leaves alternate. Blade thin leathery, lustrous adaxially, reduced green in mosaic pattern, ovate to ovate-elliptic, apex acuminate and drooping, entire and undulate, dense lateral veins.

備註 Remarks

本樹木學名根據《華南植物園導賞圖鑑》。

Scientific name of this tree is based on《華南植物園導賞圖鑑》.

植物是自養生物。它們不需從其他生物上獲取能量，只需在太陽底下安然地躺着，接受着陽光的洗禮。葉綠體是一個細小且對葉細胞來說不可或缺的植物細胞器官。葉綠體讓植物透過光合作用和呼吸作用為植株生產食物。白天時，它利用太陽光及空氣中的二氧化碳製作葡萄糖，我們稱此過程為光合作用。入夜後，缺乏光照的植物會暫緩光合作用，呼吸作用如白天時如常運作，故植物會使用氧氣分解日間所製造的大量葡萄糖以獲取能量（三磷酸腺苷）供植物所需。事實上，在光合作用過程中，葉綠體中綠色色素 —— 葉綠素只能吸收藍色和紅色波長的陽光，而綠色波長會反射至我們的眼睛。這解釋了我們會覺得葉片是綠色的原因。

葉綠體讓植物擁有生產力，但同時亦增加植物面臨被食草動物吞噬的風險。為了減低被食草動物吞食的風險，部分植物減少了葉片中的葉綠素含量，令葉片變成雜色。換言之，他們為了不「亮麗」而犧牲了生產食物的能力。有趣的是，雖然雜色的葉片確實有助於它們擺脫食草動物，但卻激起了喜歡觀葉植物的人類將植物培育作各種雜色圖案的欲望。

Plants are autotrophs. They do not need to obtain energy from other organisms, but rather layup peacefully under the sun. Chloroplast as a tiny but dominant organelle in plant cells, especially leaf cells, offers plants the ability to produce their own food through "photosynthesis" and "respiration". During the daytime, it utilizes the sunlight energy for manufacturing glucose; here is what we call photosynthesis. During the nighttime, it breaks down the glucose to obtain energy (adenosine triphosphate, ATP) that the plants need; here we refer the process to respiration. For the reason that only blue and red wavelengths of sunlight can be absorbed by chlorophyll (a type of green pigment in chloroplast) during photosynthesis, green wavelength is reflected back to our eyes and explains why leaves are green.

Chloroplast not only renders plants the productivity, but also exposes them to the risk of herbivores. To escape from herbivores, some plants reduce their chlorophyll contents in leaves and become variegated. In other words, they sacrifice the food producing ability for not being too "vivid". Although this leaf pattern does help them to get rid from herbivores, ironically, it turns out favourable to humans who intentionally cultivate plants into various foliar variegation patterns for ornamental interests.

生命力 VITALITY

花葉垂葉榕對陽光、濕度和溫度要求不高。然而，它的生長速度極快，其淺根習性使根部廣泛延伸，故在開揚廣闊的地方種植時應考慮會否損害地下公用管道設施，如電纜、光纖、水管等。

The tree shows a low demand to sunlight, humidity and temperature. However, its shallow roots are extensive and grow dramatically fast. Therefore, when planting in open areas, it should be considered whether it will damage underground public utility facility, such as cables, optical fibers, water pipes, etc.

黃榕 又稱：黃金榕

Golden Chinese Banyan | *Ficus microcarpa* L.f. 'Golden Leaf'

相片拍攝地點：海泓道、沙田公園
Tree Location: Hoi Wang Road, Sha Tin Park

名字由來 MEANINGS OF NAME

屬名 *Ficus* 源自拉丁詞 *fici*，意指無花果，是榕屬獨有的果實類型。種加詞 *microcarpa* 意指果實細小。栽培種加詞 Golden Leaf 形容其遺傳至細葉榕的金綠色葉片。

The generic name *Ficus* is derived from the Latin word *fici* (fig), referring to the emblematic fruit type of this genus. The specific epithet *microcarpa* means small-fruited. The cultivator epithet 'Golden Leaf' describes its golden-green foliage which is a cultivated characteristic distinct from the wild *Ficus microcarpa* (Chinese Banyan).

本地分佈狀態 DISTRIBUTIONS	外來物種 Exotic species
原產地 ORIGIN	野生黃榕原生於南中國、南亞、東南亞、太平洋島嶼、澳洲東部和琉球群島。 Its wild type is native to Southern China, South Asia, Southeast Asia, Pacific Islands, Eastern Australia and the Ryukyu Islands.
生長習性 GROWING HABIT	常綠喬木或灌木。高度可達 12 米。 Evergreen tree or shrub. Up to 12 m tall.

花果期 月份	1	2	3	4	5	6	7	8	9	10	11	12

花期：不詳。果期：不詳。
Flowering period: Unknown. Fruiting period: Unknown.

① 樹幹 TRUNK	② 樹皮 BARK	③ 葉 LEAVES
④ 花 FLOWERS	⑤ 果 FRUITS	

① 氣根從老枝垂下,到達地面後轉為木質。

Aerial roots hanging down from old branches, eventually lignified after reaching the ground.

② 樹皮光滑,淡灰色,具皮孔。小枝淡紅色,具葉痕。

Bark smooth, light grey, lenticellate. Branchlets pale red, with leaf scars.

③ 單葉互生。葉片革質,無毛,橢圓形,嫩葉常呈亮金黃色,卵狀橢圓形或倒卵形,基部楔形,頂端鈍形,葉緣全緣,具基脈3 條,於葉緣處連接成邊脈 / 緣內脈。托葉披針形,早落。

Simple leaves alternate. Blade leathery, glabrous, elliptic, juvenile leaves more golden, ovate-elliptic or obovate, base cuneate, apex obtuse, margin entire , basal veins 3, connecting into intramarginal veins near margin. Stipules lanceolate, caducous.

④ 雌雄同株。隱頭花序，腋生，近球形，無梗，呈綠色。

Monecious. Hypanthodia axillary, subglobose, sessile, green.

⑤ 隱頭果，成熟時轉為紅色。

Syconia turning red when mature.

應用 APPLICATION

　　黃榕作為絕佳的觀賞樹種能提供多種用途。它既可單獨種植，亦可與其他植物一併種植。黃榕與細葉榕相比，黃榕的樹形如柳樹般修長，而且擁有金燦燦的葉片，故常替代細葉榕種植在小型公園和花園內。黃榕耐修剪，常被修剪成球狀，用作園藝造景和樹籬。雖然黃榕能被修剪成獨特造型，但其生長速度快，需恆常修剪以維持形狀。

As an eminent ornamental species, Golden Chinese Banyan is valued for its versatility. It can be grown either in solitary or mix-planting. It is always planted in small parks and gardens as a surrogate of Chinese Banyan due to its relatively willowy tree form and evident golden foliage. Since the tree is hardy to pruning, it is commonly shaped into a spherical appearance, serving as topiaries and hedges. Although it can be pruned into unique shapes, its fast growth requires heavy pruning practice to maintain the tree in a decent shape.

生命力 VITALITY

　　黃榕遺傳了細葉榕強大的適應力，在瞬息萬變的環境下仍能茁壯成長。其對強風、鹽分、乾旱和污染物具高耐受性，故被認為是代替細葉榕種植在路邊和公園的絕佳替代品。

Inheriting the vigour from Chinese Banyan, Golden Chinese Banyan also shows a strong resilience to dynamic environments. It can tolerate strong wind, salt, drought and pollutants, hence considered as a great alternative for roadsides and parks.

備註 Remarks
本樹木學名根據《華南植物園導賞圖鑑》。
Scientific name of this tree is based on 《華南植物園導賞圖鑑》.

聚果榕 又稱：優曇華

Cluster Fig, Gular Fig, Country Fig, Udumbara | *Ficus racemosa* Linnaeus

相片拍攝地點：寶琳
Tree Location: Po Lam

名字由來 MEANINGS OF NAME

　　俗名 Cluster Fig 及種加詞 *racemosa* 意指聚果榕的榕果常以總狀花序簇生於小枝上。

　　聚果榕的花序為隱頭花序，其花朵匿藏在巨大的複合花托內。古人對聚果榕的生殖生物學的認知不足，或存在誤解，他們認為聚果榕的花期曇花一現，因此將此樹命名為「優曇華」。優曇華（梵語：उडुम्बर）是一種開花植物，記載在佛教最具代表性的經文《法華經》中：此花每隔 3000 年才會開花。唐代的《一切經音義》亦有描述這樹種：其果實大小可辨，開花隱蔽。

本地分佈狀態 DISTRIBUTIONS	外來物種 Exotic species
原產地 ORIGIN	東亞洲南部、印度和澳洲。 Southern East Asia, India and Australia.
生長習性 GROWING HABIT	常綠喬木。高度可達 30 米。 Evergreen tree. Up to 30m tall.

1	2	3	4	5	6	7	8	9	10	11	12

花果期
月份

花期：本港五月至七月。果期：本港五月至七月。
Flowering period: May to July in Hong Kong. Fruiting period: May to July in Hong Kong.

The common name "Cluster Fig" and the specific epithet *racemosa* describes its figs always clustered racemosely on branchlets.

The "flowers" of the tree are always confused with its hypanthodia, a type of inflorescence which the true flowers are packed in an enlarged compound receptacle. By reason of the misunderstanding of its reproductive biology, people in the past used to believe that the flower was transient and rare, hence naming the tree as "Udumbara". Udumbara (Sanskrit: उडुम्बर) is a flowering plant depicted in the *Lotus Sutra*, one of the most representative scriptures of Buddhism; the tree was described as which each of its blossoms occurs after every 3000 years. The tree was also mentioned in the *Yiqiejing yinyi* published during the Tang Dynasty for appreciating its observable fruit size and cryptic flowering.

辨認特徵 TRAITS FOR IDENTIFICATION

① 樹幹 TRUNK	② 樹皮 BARK	③ 葉 LEAVES
④ 花 FLOWERS	⑤ 果 FRUITS	

① 成熟時具板根，沒有氣根。

Buttressed when mature, aerial root absent.

② 樹皮灰棕色，光滑。

Bark greyish brown, smooth.

③ 托葉卵狀披針形，被短柔毛。單葉互生，幼時被白色柔毛。葉片革質，橢圓狀倒卵形、橢圓形或狹橢圓形，基部楔形至鈍形，頂端漸尖至鈍形，葉緣全緣，葉面呈暗綠色，葉背呈淡綠色。

Stipules ovate-lanceolate, pubescent. Simple leaf alternate, covered with white soft pubescence when young. Blade leathery, elliptic-obovate, elliptic, or narrowly elliptic, base cuneate to obtuse, apex acuminate to obtuse, entire, dark green adaxially, pale green abaxially.

④ 隱頭花序，橢圓狀卵形，聚生於老莖的小枝上。雌雄同株。

Hypanthodia ellipsoid-ovoid, aggregated on short branchlets of old stem. Monoecious.

⑤ 隱頭果，成熟時轉為紅橙色。

Syconia, turning reddish orange when mature.

應用 APPLICATION

　　與其他榕屬植物相比，聚果榕在香港較少種植。然而，聚果榕廣泛分佈在印度的森林和發展用地。其葉片、果實和根部具藥效，故常被當地人採摘。根部可用於治療糖尿病和痢疾，樹皮可以紓緩泌尿和皮膚疾病；葉片萃取物具有極佳的抗炎效用，可以緩解由組織胺和血清素引起的綜合症。

Compared with other *Ficus* spp., Cluster Fig is relatively obscure to plant in Hong Kong. The tree, however, grows throughout the forests and exploited areas in India. To the locals, the tree is always harvested for medicinal uses with its leaves, fruits and roots. For example, the roots are used for treating diabetes and dysentery. The bark can relieve urinary and skin diseases. The leaf extracts show excellent anti-inflammatory activity to relieve the syndromes induced by histamine and serotonin.

植物趣聞 ANECDOTE ON PLANTS

青果榕與聚果榕 Common Red-stem Fig and Cluster Fig：

　　青果榕與聚果榕在樹形上完全一致，它們的果實皆簇生於小枝上。要區分兩者，可從樹皮和葉形入手。聚果榕的樹皮經常裂開成薄片狀，而青果榕樹皮光滑，甚少裂開或呈薄片裂開。聚果榕的葉片呈橢圓狀倒卵形，基部楔形至鈍形，而青果榕的葉片則呈寬卵形至卵狀橢圓形，基部圓形至淺心形。

Two trees are morphologically identical, both characterising the clustered fruits on branchlets. To distinguish the trees, the trunk of Cluster Fig is always covered with flakes, whereas the one of Common Red-stem Fig is smooth. Different from the leaves of Cluster Fig that are elliptic-obovate, with cuneate to obtuse base, those of Common Red-stem Fig are broadly ovate to ovate-elliptic, with rounded to shallowly cordate base.

備註 Remarks

本樹木學名根據中國植物誌網頁：

Scientific name of this tree is based on the Flora of China website：

http://www.efloras.org

心葉榕 又稱：假菩提樹
Mock Peepul Tree, Mock Bodh Tree | *Ficus rumphii* Blume

相片拍攝地點：維多利亞公園、旺角大球場
Tree Location: Victoria Park, Mong Kok Stadium

名字由來 MEANINGS OF NAME

此品種的中文名稱「心葉榕」，源自其心型葉片。由於心葉榕與菩提樹外形相似，故亦被稱為假菩提樹。

The tree is named as「心葉榕」in Chinese, describing its heart-shaped leaves. Considering its indistinguishable appearance to *Ficus religiosa (Peepul Tree)*, *it is also named as "Mock Peepul Tree" while* mock means "pretending somethings".

應用 APPLICATION

心葉榕含豐富藥用價值。其樹皮可治療由蛇咬引起的血尿、痕癢或白斑。此外，將心葉榕的乳汁與胡椒和薑黃混合並服用便能驅除體內寄生蟲。乳汁還具有止咳、抗利尿和催吐等醫藥作用。

本地分佈狀態 DISTRIBUTIONS	外來物種 Exotic species
原產地 ORIGIN	心葉榕分佈在中國中部及南部。此外亦分佈在東南亞國家，包括印度、泰國、馬來西亞、尼泊爾和越南。 The tree is distributed in South-Central China. It is also spread throughout the countries of Southeast Asia, including India, Thailand, Malaysia, Nepal and Vietnam.
生長習性 GROWING HABIT	半落葉喬木。高度可達 15 米。 Semi-deciduous tree. Up to 15 m tall.

花果期月份	1	2	3	4	5	6	7	8	9	10	11	12

花期：本港五月至九月。果期：本港五月至九月。
Flowering period: May to September in Hong Kong. Fruiting period: May to September in Hong Kong.

Mock Peepul Tree is valued for its versatile medicinal functions. Its bark can treat haematuria, itching and leucoderma caused by snake bite. Additionally, combining the latex with pepper and turmeric can expel worms in the human body; the solitary application of latex also shows wonderful anticough, antidiuretic and emetic effects.

辨認特徵 TRAITS FOR IDENTIFICATION

① 樹幹 TRUNK	② 樹皮 BARK	③ 葉 LEAVES
④ 花 FLOWERS	⑤ 果 FRUITS	

① 具板根，甚少具氣生根。

Buttressed when mature, rarely with aerial root.

② 樹皮光滑，灰綠色。

Bark smooth, greyish green.

③ 單葉互生，革質，心型至卵狀心型，基部心型至寬楔形，頂端漸尖或短尾形，無毛。基脈 5 條，最外 2 條基脈較短。托葉卵狀披針形，葉痕明顯。

Simple leaves alternate. Blade leathery, cordate to ovate-cordate, base cordate to broadly cuneate, apex acuminate or shortly caudate, glabrous. Basal veins 5, outer 2 basal veins short. Stipules ovate-lanceolate, leaf scar predominant.

④ 雌雄同株，隱頭花序，腋生，球狀，無柄。

Monoecious. Hypanthodium, axillary, globose, sessile.

⑤ 無花果，初時滿佈圓點，成熟時由綠色轉為紫黑色。

Syconium globose, initially full of dots, turning from green to purple black when mature.

心葉榕與菩提樹的差別 Mock Peepul Tree and *Ficus religiosa* (Peepul Tree)：

心葉榕與菩提樹皆屬於桑科，它們的外觀如出一轍。以下將會介紹如何區分「真」「假」菩提樹：首先，心葉榕的葉片較短，大約 6 至 13 厘米，而菩提樹的葉片較長，為 9 至 17 厘米。此外，心葉榕的葉柄較短，約 6 至 8 厘米，而菩提樹的葉柄則約長 6.5 至 13 厘米。心葉榕的葉尖稍短。相反，菩提樹的葉尖較長，約 2 到 5 厘米，幾乎是整片葉片長度的一半。最容易分辨兩者的地方為樹皮，心葉榕的樹皮呈灰綠色，而菩提樹則為紅棕色。

Being the members of Moraceae, two trees are always confused with their similar morphologies. The following introduces how to distinguish between the "true" and "fake" Peepul tree: first, the leaves of Mock Peepul Tree are relatively shorter, around 6 to 13 cm, while that of Peepul Tree are longer, with 9 to 17 cm. In addition, Mock Peepul Tree has shorter petioles at 6 to 8 cm, while the petioles of Peepul Tree are about 6.5 to 13 cm. The leaf apex of Mock Peepul Tree is slightly shorter. On the contrary, the one of Peepul Tree is notably longer (about 2 to 5cm) and attains to almost half of the blade length. The easiest way to tell the two apart is by comparing their barks: the "fake" one is greyish green while the "real" one is reddish-brown.

古樹名木 Old and Valuable Trees (OVT)：

旺角大球場內的心葉榕（編號：LCSD YTM/ 107）是目前唯一一棵心葉榕被列入香港古樹名木，其胸徑約為 1030 毫米，高 11 米，樹冠寬 24 米。此外，香港動植物公園內還有一棵心葉榕曾被列入 OVT（編號：ARCHSD CW/27），可惜由於該樹感染嚴重褐根病，故於 2021 年 3 月已被移除。

古樹名木不僅為當地動物提供豐富的生態區位和食物，它們亦為這片土地留下獨有的自然地貌，故應得到大眾關注及充分保護。

The Mock Peepul Tree (No. LCSD YTM/107) in the Mongkok Stadium is currently the only tree of this species that is listed in Hong Kong's OVT. Its diameter at breast height (DBH) is about 1030 mm, its height is 11 m, and its crown spread is 24 m wide. Indeed, there was once another listed Mock Peepul Tree OVT in the Hong Kong Zoological and Botanical Gardens (ARCHSD CW/27); Unfortunately, the tree was removed in March 2021 due to the severe infection of Brown Root Rot disease.

OVT not only provide abundant niches and food for the local animals, they also leave a unique natural landscape for this land, so they should be treasured and fully protected by the public.

心葉榕古樹名木

筆管榕

Japanese Superb Fig, Superb Fig | *Ficus subpisocarpa* Gagnep.

相片拍攝地點：荔枝角公園、九龍公園
Tree Location: Lai Chi Kok Park, Kowloon Park

名字由來 MEANINGS OF NAME

　　此品種有一個有趣的中文名稱「筆管榕」和英文名稱 Pen Tube Fig，都是在描述此品種具粗壯而細長的紅葉芽。這些名稱亦暗示其葉紅色的嫩芽，由於硃砂也是紅色的，故與蘸硃砂的中國毛筆筆尖相似。

　　The tree is given with an interesting Chinese name「筆管榕」, known as "Pen Tube Fig" in English. The names describe its rather robust and elongated red leaf buds with a great similarity to the tip of a Chinese brush dipped in cinnabar, which is red.

本地分佈狀態 DISTRIBUTIONS	原生物種 Native species
原產地 ORIGIN	中國南方和西南省份，如：福建、廣東、廣西和海南。 South and Southwest China, eg. Fujiang, Guangdong, Guangxi and Hainan.
生長習性 GROWING HABIT	落葉喬木。高度可達 15 米。 Deciduous tree. Up to 15 m tall.

1	2	3	4	5	6	7	8	9	10	11	12	花果期 月份

花期：本港二月至九月。果期：本港二月至九月。
Flowering period: February to September in Hong Kong. Fruiting period: February to September in Hong Kong.

筆管榕的樹皮堅硬，可以用來製作雕塑。其根部和葉片可治療中毒、炎症、口腔念珠菌病（鵝口瘡）和漆瘡。此外，筆管榕的樹冠巨大，具優良遮蔭效果。因此，它被廣泛種植在街道或公園中。

Its bark is stiff and can be used for making sculptures. The roots and leaves can treat toxification, inflammation, oral candidiasis and lacquer dermatitis. In addition, Japanese Superb Fig has a large crown that provides excellent shading. Therefore, it is widely planted in streets or parks.

辨認特徵 TRAITS FOR IDENTIFICATION

① 樹幹 TRUNK	② 樹皮 BARK	③ 葉 LEAVES
④ 花 FLOWERS	⑤ 果 FRUITS	

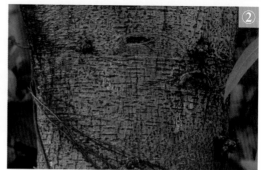

① 具板根，甚少具氣生根。

Buttressed, seldomly with aerial roots.

② 樹皮呈黑棕色，小枝呈淺紅色，無毛，受傷時分泌白色乳汁。

Bark blackish brown, branchlets pale red, glabrous, secreting milky sap when wounded.

③ 葉片簇生於枝條頂端。葉片近紙質，橢圓形至長橢圓形，基部圓形，頂端短漸尖，葉緣稍波狀或全緣。葉脈明顯，具 3 條基出脈及 7-9 對側脈。托葉披針形，被疏短柔毛。春季時紅色嫩芽着生於枝條頂端。

Stipules lanceolate, sparsely pubescent. Leaves alternate, crowded at the apex of branches. Blade subpapery, elliptic to oblong, base rounded, apex short-acuminate, slightly undulate or entire, leaf veins conspicuous, with 3 basal veins and 7-9 pairs of lateral veins. Red leaf buds germinating at the apex of branches in spring, red.

④ 雌雄同株，扁球形。榕果成對或單生腋生於小枝。不剖開的情況下無法觀察內部的雄花及雌花。

Monoecious, flat spherical, stalked, axillary at branchlets, paired or solitary. Male and female flowers unobservable without dissection.

⑤ 榕果果柄較短，帶白點，着生於無葉枝條上，成熟時轉為紫黑色或粉色。

Figs have short fruit stalks, with white spots, growing on leafless branches, syconia turning purple black or pink when ripe.

生態 ECOLOGY

筆管榕的果實（榕果）為鳥類和其他動物提供了豐富的食物來源，從而增加了本港的生物多樣性。

Its fruits serve as a wholesome delicacy to birds and other animals, and thus enhance the local biodiversity in Hong Kong.

植物趣聞 ANECDOTE ON PLANTS

筆管榕與黃葛樹的差別 Japanese Superb Fig and *Ficus Virens* (Big-leaved Fig)：

筆管榕與黃葛樹是香港市區常見樹種，它們皆被廣泛種植在公園和花園中。它們的外觀相似，如果缺乏深入觀察，便會很容易將兩者混淆。要清楚分辨兩者，可從樹皮、葉序、果柄及果實顏色入手。筆管榕的樹皮呈黑棕色，而黃葛樹的樹皮則呈黃棕色，且光滑。筆管榕的葉片簇生，而黃葛樹則為互生，葉背側脈明顯凸起。筆管榕果柄短，成熟時轉為紫黑色或粉色。但黃葛樹缺少果柄，果實成熟時呈紫紅色。

Japanese Superb Fig and Big-leaved Fig are pervasively planted in parks and gardens by virtue of their majestic horizons and dense foliage that can render an unsurpassed shading effect. On account of their similar outward appearances, the trees are always mistakenly identified. To distinguish the trees, first, the bark of Japanese Superb Fig is blackish brown, while that of Big-leaved Fig is yellowish-brown. Second, the leaves of Japanese Superb Fig are often clustered at the apex of branches, but those of Big-leaved Fig are scattered and the lateral veins are rather prominent on the beneath of leaves. The fruits of Japanese Superb Fig are shortly stalked and turn purple-black or pink at maturity, whereas those of Big-leaved Fig are sessile and turn purplish red when ripe.

桑 又稱：桑樹

White Mulberry | *Morus alba* L.

相片拍攝地點：柴灣公園、香港動植物公園
Tree Location: Chai Wan Park, Hong Kong Zoological and Botanical Gardens

名字由來 MEANINGS OF NAME

種加詞 *alba* 有白色的含意。

記載在《搜神記》中的「蠶馬」，是個關於白桑樹起源的悲劇故事。在古代，一位女子因丈夫突然失蹤而感到悲痛欲絕。她許下承諾，若有人能找到她的丈夫，那人便能娶她的女兒。然而，把她丈夫帶回來的竟是一匹馬！由於許下了承諾，她無可奈何下只好把女兒嫁給馬。女子的丈夫難以接受，於是一怒之下便殺了馬匹，再把馬的外皮放在院子外。令人始料未及的是，他們的女兒和馬的外皮其後竟一同失蹤。當村民找到他們的女兒時，她已經變成了絲綢，馬皮則在樹上變成了繭。「桑」字在漢語中與「喪」同音，因此人們將這棵樹命名為桑樹，以紀念這段令人心酸的故事。

本地分佈狀態 DISTRIBUTIONS	原生物種 Native species
原產地 ORIGIN	桑樹原生於中國中北部和中南部，並被廣泛引入至阿富汗、哥斯達黎加和墨西哥等不同國家。 The species is native to North-Central and South-Central China and widely introduced to countries as diverse as Afghanistan, Costa Rica and Mexico.
生長習性 GROWING HABIT	落葉喬木。高度可達 10 米。 Deciduous tree. Up to 10 m tall.

花果期 月份

1	2	3	4	5	6	7	8	9	10	11	12

花期：本港二月至八月。果期：本港二月至八月。
Flowering period: February to August in Hong Kong. Fruiting period: February to August in Hong Kong.

The specific epithet *alba* means white.

Silk Horse（蠶馬）is a tragic story written in *Anecdotes About Spirits and Immortals*（搜神記）, a legend talking about the origin of White Mulberry tree. In ancient times, a woman was pathetic about the abrupt disappearance of her husband. To seek her husband, she permitted anyone who found her husband could marry her daughter. One day, someone finally brought her husband back, while someone referred to a horse! Under the marriage commitment, her daughter was compelled to marry the horse. Such absurd news stunned her husband and he killed the horse with anger, then put the horse's skin outside the courtyard. Unbelievably, their daughter disappeared with the horse's skin. When villagers found their daughter, she had already transformed into silk and the skin of the horse had converted into a cocoon hanging on the tree. 「桑」(Mulberry) is a homophone of 「喪」(mourning) in Chinese; people thus named the tree as "Mulberry" to commemorate this heart-wrenching story.

應用 APPLICATION

桑樹樹皮由高纖維組成，通常用以製造人造絲或紙張。桑樹紙張堅韌耐用。它在古代更是中國畫紙的製作原料。桑樹葉片具藥用價值，在傳統中醫學中用於治療發燒、悶熱和頭暈。它的葉片還具有抗菌、抗高血壓和抗高血脂的作用。其果實可食用，亦可加工成酒和果醬，種子還可以提煉成植物油。

The bark is composed of high fibre which is favourable for making rayons and papers. Mulberry paper is tough and durable, hence highly recommended as a material of the traditional Chinese painting paper in ancient times. The leaves serve the primary ingredients of traditional Chinese medicines which can treat fever, muggy and dizziness. The leaves also exhibit antibacterial, antihypertensive and anti-hyperlipidemic effects. The fruits can be eaten fresh but is more often to be processed into wine and jam. Its seeds can also be refined into vegetable oil.

辨認特徵 TRAITS FOR IDENTIFICATION

① 樹幹 TRUNK	② 樹皮 BARK	③ 葉 LEAVES
④ 花 FLOWERS	⑤ 果 FRUITS	

① 枝條平滑，具皮孔，圓柱狀。
Branchlets lenticellate, terete, finely hairy.

② 樹皮粗糙，灰色，具淺溝。
Bark greyish colour, coarse, shallowly furrowed.

③ 單葉互生，紙質，葉形多變，卵形至寬卵形，有時具不同開裂，基部圓形至心形，頂端銳形、漸尖或鈍形，葉緣具粗鋸齒至鈍鋸齒，基脈 3-5 條。

Simple leaves alternate. Blade papery, shapes largely variable, ovate to broadly ovate, serrate, sometimes lobed, base round to cordate, apex acute, acuminate or obtuse, margin coarsely serrate to crenate, basal veins 3-5.

④ 雌雄同株異花，單性花，葇荑花序，腋生。花朵較小，呈綠色，無柄。雄花序下垂，密被白色長柔毛。雌花具柱頭 2 深裂。

Monoecious. Catkins axillary. Flowers small, green, sessile. Male inflorescence pendulous, densely white villose. Female flowers with stigma 2-parted.

⑤ 聚花果，眾多瘦果果實，肉質，卵狀橢圓形或圓柱狀，初時綠色或紅色，成熟時轉為深紫色。

Multiple fruit, with many achenes, fleshy, ovoid-ellipsoid or cylindric, red or dark purple when ripe.

生態 ECOLOGY

蠶蟲以桑葉為食。當蠶蟲吐絲結繭後，蠶繭經抽絲剝繭及紡織，便成為絲綢。

Silkworms feed on mulberry leaves. After the silkworm spins the cocoons, the cocoons are stripped and spun to become silk.

生命力 VITALITY

桑樹具有豐富的經濟價值，並在世界各地廣泛種植，用以生產絲綢和飼料。桑樹對不同類型的土壤、乾旱和鹽分具高耐受性。同時，它具有快速生長的習性，故在部分國家，如在巴西和美國俄勒岡州等，被視為入侵品種。

The Mulberry Tree is rich in economic value and is widely cultivated all over the world for the production of silk and fodder. It is highly resilient to different types of soil, drought and salinity. With its rapid growth rate, it is regarded as a notorious invasive species in some countries, such as Brazil and Oregon in the United States.

黧蒴錐 又稱：裂斗錐栗

Chestnut Oak, Castanopsis | *Castanopsis fissa* (Champ. ex Benth.)
Rehder & E. H. Wilson

相片拍攝地點：元朗公園、迪欣湖
Tree Location: Yuen Long Park, Inspiration Lake Recreation Centre

名字由來 MEANINGS OF NAME

種加詞 *fissa* 的意思是裂隙，是指其果實成熟時會分裂成不規則瓣片。

The specific epithet *fissa* means fissure, referring to its fruits splitting into segments at maturity.

應用 APPLICATION

黧蒴錐是一種生長迅速的常綠喬木，對貧瘠土壤具高耐受性。它是香港早期造林計劃中為數不多成功扎根的原生物種，故一直被選為改善水土流失的先鋒品種。

黧蒴錐木材細膩，是製作家具、門和箱板的好材料。其果實含豐富澱粉，據說在日治時期，一些香港原居民在饑荒時會採集了這些果實來熬粥。

本地分佈狀態 DISTRIBUTIONS	原生物種 Native species
原產地 ORIGIN	原生於中國東南部各省份和香港，亦分佈於泰國、越南和老撾。 Native to Southeast China and Hong Kong, it also distributed in Thailand, Vietnam and Laos.
生長習性 GROWING HABIT	大型常綠喬木。高度可達 20 米。 Large evergreen tree. Up to 20 m tall.

1	2	3	4	5	6	7	8	9	10	11	12

花果期月份

花期：本港四月至六月。果期：本港十月至十二月。
Flowering period: April to June in Hong Kong. Fruiting period: October to December in Hong Kong.

It is a fast-growing evergreen tree with great resilience to infertile soils. It is one of the few native species that earned a success during the early reforestation in Hong Kong. It has been still engaged as a pioneer tree species to ameliorate soil erosion.

The wood is fine and favourable for making furniture, doors and box boards. The profusely starchy fruits were once collected and processed into congee by Hong Kong villagers when there was starvation during the Japanese occupation.

辨認特徵 TRAITS FOR IDENTIFICATION

① 樹幹 TRUNK	② 樹皮 BARK	③ 葉 LEAVES
④ 花 FLOWERS	⑤ 果 FRUITS	

① 鰲蕭錐的樹幹。

Trunk of *Castanopsis fissa* (Champ. ex Benth.) Rehder & E. H. Wilson.

② 樹皮灰棕色，成熟時粗糙。小枝紅紫色，棱明顯。

Bark greyish brown, coarse when mature. Branchlets red-purple, ribs conspicuous.

③ 葉片互生，厚紙質，形狀大小多變，長橢圓形至倒卵狀橢圓形，基部楔形，頂端短漸尖至圓形，葉緣波狀，葉片下半部分具圓齒。側脈眾多，通常 15-20 對，葉背側脈凸起。葉面無毛，幼時葉背被黃棕色柔毛及銹色毛狀體／細伏毛，成熟時近無毛。

Leaves blade alternate, thick papery, highly variable in size and shape, oblong to obovate-elliptic, base cuneate, with rounded teeth, undulate and crenate on the lower half. Lateral veins many, usually 15-20 pairs, raised abaxially. Adaxially glabrous, abaxially yellowish brown puberulent first, glabrescent.

④ 穗狀花序直立，組成圓錐花序，似煙火。
單性花，雌雄同株，雄花眾多，雄蕊呈白色，
擠壓／簇生。

Spikes erect, forming a panicle, resembling fireworks.
Unisexual, monoecious, with many male flowers,
stamens white, compressed/clustered.

⑤ 殼斗卵球形至橢球形，被深紅棕色絨毛，
未成熟時堅果被殼斗完全包裹，成熟時不
規則 2-3 瓣裂。堅果種子一粒，球形至橢圓
形，呈棕紅色。

Cupule ovoid to ellipsoid, slightly dark reddish
brown tomentose, fully enclosing the nut when
immature, splitting into 2-3 irregular segments
at maturity. Nut 1 per cupule, globose to elliptic,
brown red.

生態 ECOLOGY

　　黧蒴錐的殼斗難以咀嚼，通常只對嚙齒動物及其他分散囤積動物具吸引力。由
於缺乏種子掠食者，殼斗只能依靠重力把種子散播出去，導致同一範圍過度擠擁或種
子分佈過於密集。有研究指出，森林中的黧蒴錐種子傳播距離通常限制在 5 米以內。

The cupule of Chestnut Oak is extremely hard to chew and attracts only rodents and some scatter
hoarders. Due to the absence of seed predators, the cupule can only be dispersed by gravity and
crammed in a close distance. According to some studies, the distance of seed dispersal of Chestnut Oak is
probably confined to only 5 m in forests.

簕杜鵑 又稱：葉子花、毛寶巾、九重葛

Brazil Bougainvillea, Beautiful Bougainvillea | *Bougainvillea spectabilis* Willd.

相片拍攝地點：柴灣公園
Tree Location: Chai Wan Park

本地分佈狀態 DISTRIBUTIONS	外來物種 Exotic species
原產地 ORIGIN	巴西。 Brazil.
生長習性 GROWING HABIT	攀援灌木。莖長達 10 米或以上。 Scandent shrub. Up to 10 m or more tall.

花果期 月份	1	2	3	4	5	6	7	8	9	10	11	12

花期：本港全年可見。果期：不詳。
Flowering period: Visible throughout the year in Hong Kong. Fruiting period: Unknown.

為了紀念首位於 1768 年在巴西發現葉子花屬的法國軍官 — 路易・安托萬・德・布干維爾（1729-1811），故取其名作屬名 *Bougainvillea*。

雖然其中文名為「簕杜鵑」，但事實上與杜鵑花科的杜鵑花屬相差甚遠。相反地，「簕杜鵑」形容其花朵如杜鵑花般嬌艷，而小枝多刺。

The generic name *Bougainvillea* commemorates Louis Antoine de Bougainville (1729-1811), a French army officer who firstly discovered the genus *Bougainvillea* in Brazil in 1768.

Although it is named as 「簕杜鵑」 in Chinese, it is phylogenetically far from *Rhododendron* in Ericaceae. Instead, 「簕杜鵑」 depicts which the flowers are as charming as Rhododendron while the branchlets are spiny.

簕杜鵑壯麗迷人，而且花期甚長，故被視為高觀賞性的園藝植物。值得一提的是，我們眼見華麗奪目的部分並非花瓣，而是苞片，其花瓣合生且不顯眼。簕杜鵑可修剪性高，其枝條可彎曲並下垂，故被視為觀賞價值高的園藝植物種植在公園、花園，亦常被修剪成單一主莖樹木、欄柵和拱門。由於簕杜鵑生長迅速，故需經常修剪以保持理想造型。

在香港或華南地區近全年開花但從不結果，其栽培方法通常可透過扦插繁殖。扦插繁殖是一種營養繁殖，透過母莖生根發芽長出新株的無性繁殖。由於沒有授粉過程，新株與母莖的基因必然是一致的。基於以上原因，我們假設簕杜鵑的種群較預想中的狹窄，可能與染井吉野櫻有相似結果，即是所有獨立個體遺傳自同一株母株（詳見請參考鐘花櫻桃）。

Brazil Bougainvillea is valued for its glamorous flowers that are long blooming in Hong Kong. Remarkably, the rather magnificent and colourful parts of the flower are regarded as bracts but not petals, which are fused and inconspicuous. Since Brazil Bougainvillea is highly malleable with rather bendable and pendulous branches, it is a versatile ornamental component for parks and gardens, frequently planted into a solitary tree, fences and arches. The plant is relatively fast-growing and requires heavy pruning work to maintain a decent shape.

In Hong Kong or southern China, Brazil Bougainvillea blooms all year round but is hard to produce seeds and the cultivation of which is mainly through stem cutting, a kind of vegetative reproduction that a new individual sprouts up and roots from a separated mother stem or branch. Since there is an absence of mating, the individual must be genetically the same with the mother stem. Based on it, we assume that the population of Brazil Bougainvillea could be much narrower than expected in Hong Kong, probably sharing the similar phenomenon as *Prunus × yedoensis* that the genetic source of the individuals is limited to one mother tree (please refers to *Prunus campanulata* for more details).

① 樹幹 TRUNK　　② 樹皮 BARK　　③ 葉 LEAVES
④ 花 FLOWERS

① 簕杜鵑的樹幹。

Trunk of *Bougainvillea spectabilis* Willd.

② 莖皮灰棕色。小枝被微柔毛,具皮孔,常具刺。

Stem greyish brown. Branchlets puberulous, lenticellate, often spiny.

③ 單葉互生。葉片紙質,兩面皆被微柔毛,卵形至卵狀披針形或橢圓形,基部寬楔形至圓形或截形,頂端漸尖,葉緣全緣。

Simple leaves alternate. Blade papery, puberulent on both surfaces, ovate to ovate-lanceolate or elliptic, base broadly cuneate to rounded or truncate, apex acuminate, margin entire.

④ 雌雄同株。花朵常 3 朵簇生於苞片連接處,並排列成頂生圓錐花序。苞片 3 片,葉狀,宿存,顏色多變,常呈白色、粉紅色、紫色和橙色。花被管狀,具 5 裂。

Hermaphroditic. Flowers always in cluster of 3 at the junction of bracts, terminal panicles. Bracts 3, leafy, persistent, colours multiple, commonly white, pink, purple and orange. Perianth tubular, 5-lobed.

註:本樹另有瘦果,長橢圓狀橢圓體,密生毛,呈黑色。

Remarks: Achenes oblong-ellipsoid, hairy, black.

生命力 VITALITY

種植簕杜鵑需要全日照的環境及排水良好的土壤。

Full sunlight and well-drained soils are needed for the planting.

第倫桃 又稱：五椏果
Elephant Apple | *Dillenia indica* L.

相片拍攝地點：荃灣公園
Tree Location: Tsuen Wan Park

名字由來 MEANINGS OF NAME

種加詞 *indica* 暗示第倫桃原生於印度。因其果實形狀與大象的腳趾相似，故又名「大象林檎」。

The specific epithet *indica* indicates its origin from India. By reason of its similar fruit shape to elephant's toes, it is also named as "Elephant Apple".

應用 APPLICATION

第倫桃的木材堅韌，可用作建造船舶、木箱及卡板的用料。其果實可生食，味似未熟透的蘋果，酸甜之中帶澀，常用於製作成沙律、咖喱或果醬。第倫桃果亦可減輕脫髮問題。將第倫桃的葉片、樹皮和果實混合，可用於治療癌症及腹瀉。

本地分佈狀態 DISTRIBUTIONS	外來物種 Exotic species
原產地 ORIGIN	中國中南部和東南部、印度、中南半島、馬來西亞半島、蘇門答臘、爪哇島、婆羅洲和新加坡。 South-Central and Southeast China, India, Indochina, Peninsular Malaysia, Sumatra, Java, Borneo and Singapore.
生長習性 GROWING HABIT	半落葉喬木。高度可達 30 米。 Semi-deciduous tree. Up to 30 m tall.

| 1 | 2 | 3 | 4 | 5 | 6 | 7 | 8 | 9 | 10 | 11 | 12 | 花果期月份 |

花期：本港四月至五月。果期：本港十月至十二月。
Flowering period: April to May in Hong Kong. Fruiting period: October to December in Hong Kong.

第倫桃的葉片茂密，樹形優美，故被視為觀賞樹種種植在公園和花園中，為途人提供優良的遮蔭效果。作為觀賞樹種，在花期吸引蜜蜂和蝴蝶勤勞地採蜜，形成如畫的風景，為城市增添了生命力。然而，它的果實出乎意料地沉重，故應在其成熟前摘下，以確保公眾安全。

The wood is stiff and demanded for making ships, boxes and pallets. In addition, the fruits are esculent and taste like unripe apples. Other than eating it fresh, the fruits can also be chopped into pieces to embellish salads or processed into curries and jam. The fruits are wholesome and can support daily nutrients to avoid hair loss.

The tree is valued as an excellent ornamental tree due to its handsome tree form, dense foliage and majestic blossoms. Planting as a shading tree, it is competent to provide an unsurpassed shading effect. Plant as an ornamental tree, the spectacular blossoms are assiduously tapped by bees and butterflies during the flowering seasons, providing a picturesque outlook and embellishing the city with perceived vitality. Although the fruits are appealing with a gigantic size, they are exceptionally weighty and could compromise the safety of visitors if they are not removed properly before ripening.

辨認特徵 TRAITS FOR IDENTIFICATION

① 樹幹 TRUNK	② 樹皮 BARK	③ 葉 LEAVES
④ 花 FLOWERS	⑤ 果 FRUITS	

① 第倫桃的樹幹。

Trunk of *Dillenia indica* L.

② 樹皮紅棕色，片狀剝落。幼枝被褐色短柔毛或漸變無毛，具明顯脫葉痕。

Bark reddish brown, peeling off in flakes. Young branchlets brown pubescent, glabrescent, with obvious leaf scars.

③ 單葉互生，簇生於小枝頂端。大型葉片，革質，長橢圓形或倒卵狀長橢圓形，頂端微凸，基部寬楔形，葉緣具明顯鋸齒，側脈明顯，並於葉背隆起。

Simple leaf alternate, clustered at the apex of branchlets. Blade leathery, large, oblong or obovate-oblong, apex mucronate, base broadly cuneate, coarsely serrate, lateral veins eminent abaxially.

④ 雌雄同體。大型花朵，單生花。萼片 5 片，微圓形，厚肉質。花瓣 5 片，倒卵形，呈白色，雄蕊排列成 2 輪，外輪雄蕊眾多，呈橙黃色，柱頭輪狀，呈白色。

Hermaphrodite. Flowers large, solitary. Sepals 5, approximately rounded, thickly fleshy. Petals 5, obovate, white, stamens arranged in 2 distinct groups, outer numerous, orange-yellow, stigma radical, white.

⑤ 聚合果，球形，不開裂，具萼片宿存，肥厚並包覆果實，成熟時由黃綠色轉為黃棕色。

Aggregate fruit, globose, indehiscent, baccate, persistent sepals enlarged and enclosed the entire fruit, turning yellowish green to yellowish brown at maturity.

生態 ECOLOGY

第倫桃的花朵呈白色，芬芳馥郁。花期時會被數以萬計的授粉者包圍。

Dillenia indica has fragrant white flowers. It is surrounded by tens of thousands of pollinators at flowering time.

生命力 VITALITY

第倫桃偏愛全日照、潮濕及排水良好的土壤。

The tree prefers full sunlight, moist and well-drained soils.

大頭茶 又稱：南投大頭茶
Hong Kong Gordonia | *Polyspora axillaris* (Roxb. ex Ker Gawl.)
Sweet

相片拍攝地點：針山
Tree Location: Needle Hill

本地分佈狀態 DISTRIBUTIONS	原生物種 Native species
原產地 ORIGIN	分佈於廣東、海南等華南省份。它在香港郊野公園屢見不鮮。 Guangdong, Hainan and South China. It is common in Hong Kong country parks.
生長習性 GROWING HABIT	常綠喬木。高度可達 8 米。 Evergreen tree. Up to 8 m tall.

花果期 月份	1	2	3	4	5	6	7	8	9	10	11	12

花期：本港九月至十月。果期：本港十一月至十二月。
Flowering period: September to October in Hong Kong. Fruiting period: November to December in Hong Kong.

① 樹幹 TRUNK	② 樹皮 BARK	③ 葉 LEAVES
④ 花 FLOWERS	⑤ 果 FRUITS	

① 大頭茶的樹幹。小枝粗壯，密被短柔毛。

Trunk of *Polyspora axillaris* (Roxb. ex Ker Gawl.) Sweet. Branchlets stout, densely pubescent.

② 樹皮橙棕色，光滑。

Bark orange brown, smooth.

③ 單葉互生，葉片成熟時革質。葉面暗淡無毛，葉背淡綠色，具明顯的中脈，長圓狀卵形到倒披針形，先端圓形或鈍，基部楔形和下延，全緣或在先端具圓齒，側脈模糊。

Simple leaves alternate. Blade leathery, adaxially dull and glabrous, abaxially pale green, with eminent midvein, oblong-ovate to oblanceolate, apex rounded or obtuse, base cuneate and decurrent, entire or remotely crenate on apex, lateral veins vague.

④ 單生花或成對，大小可辨。花瓣 5 片，呈白色。雄蕊眾多，呈黃色。

Flowers solitary or in pairs, axillary, large. Petals 5, white. Stamens many, yellow.

⑤ 蒴果，長橢圓形或長橢圓狀卵形。
種子扁平，具翅。

Capsules oblong or oblong-ovoid. Seeds compressed, winged.

高濃度的鋁離子對大多數植物具毒性，但大頭茶卻是例外。大頭茶會把鋁離子吸收並積存在葉片和根系中，透過與氟離子產生化學反應，降低鋁離子毒性。

High concentrations of aluminum ions are toxic to most plants, but Hong Kong Gordonia is an exception. The tree can mitigate aluminium pollution by accumulating the aluminium ions in its leaves and roots through the reaction with fluoride ions.

香港林地復育 Reforestation of Hong Kong：

香港早期林地復育目標為促進活化，如增加土壤營養含量、降低土壤侵蝕。台灣相思和木麻黃等外來物種，有生長速度快、適應力強及能固定土壤的特性，與目標完美契合，因此當年在香港的山頭上廣泛種植。雖然種植外來植物能滿足綠化需求，但這些品種未能為本地野生動物提供食物來源。為鼓勵豐富本地生物多樣性，漁護署於 2009 年展開了「郊野公園植林優化計劃」，以原生物種取代外來物種。由於大頭茶對風、乾燥和貧瘠土壤具有較高的耐受性，而且在冬季能為大胡蜂提供至關重要的花粉來源，故成為應用於計劃中的樹種。除了大頭茶，其他香港原生品種如黧蒴錐和木荷亦是計劃中的關鍵樹種，以重建香港植林的自然生態。

In the early reforestation of Hong Kong, exotic species such as *Acacia confusa* (Taiwan Acacia) and *Casuarina equisetifolia* (Horsetail Tree) which are fast-growing and able to adapt and ameliorate barren environments, were the framework trees selected primarily for the promotion of rapid soil regeneration. Although they are competent to inject vigour into the lifeless land, they fail regrettably to provide a food source for native wildlife. To enrich the local biodiversity, the government launched the Country Parks Plantation Enrichment Programme in 2009 with a main aim of substituting the exotic trees with native species. Hong Kong Gordonia is one of the framework trees that are native to Hong Kong and show relatively good tolerance to wind, droughts, and infertile soils; it also serves as a vital food source for large vespids in winter. Not only Hong Kong Gordonia, but other native trees such as *Castanopsis fissa* (Castanopsis) and *Schima superba* (Schima) are also picked for re-establishing the Hong Kong natural forest.

厚皮香

Naked Anther Ternstroemia | *Ternstroemia gymnanthera* (Wight & Arn.) Bedd.

相片拍攝地點：黃泥涌樹木研習徑
Tree Location: Wong Nai Chung Tree Walk

名字由來 MEANINGS OF NAME

為了紀念「現代生物分類學之父」卡爾‧林奈（1707-1778）的學生 —— 瑞典博物學家克里斯托弗‧特恩斯特羅姆（1703-1746），故屬名取其名並冠名為 *Ternstroemia*。種加詞 *gymnanthera* 意指其外露的花藥。

本地分佈狀態 DISTRIBUTIONS	原生物種 Native species
原產地 ORIGIN	亞洲南部和東南內陸、中國中南部和東南部及香港。 South and Mainland Southeast Asia, South-Central and Southeast China, and Hong Kong.
生長習性 GROWING HABIT	常綠灌木或小喬木。高度可達 10 米或以上。 Evergreen shrub or small tree. Up to 10 m or above tall.

1	2	3	4	5	6	7	8	9	10	11	12	花果期月份

花期：本港七月至八月。果期：本港八月至十月。
Flowering period: July to August in Hong Kong. Fruiting period: August to October in Hong Kong.

The generic name *Ternstroemia* commemorates the Swedish naturalist Christopher Ternstroem (1703-1746), who was one of the students of Carl Linnaeus (1707-1778), "the father of modern taxonomy". The specific epithet *gymnanthera* means exposed anthers.

應用 APPLICATION

厚皮香的葉片茂密、枝幹分層，且耐陰性高，故受到重視。它常被剪形成球狀或種植作一排樹籬。厚皮香生長緩慢，可透過偶爾修剪以維持其造型。

Naked Anther Ternstroemia is valued for its dense foliage, tiered branches and superior shady tolerance. It is always pruned into a ball shape or planted into a row of hedges. Since Naked Anther Ternstroemia grows sluggishly, it can maintain a decent shape for a longer time.

辨認特徵 TRAITS FOR IDENTIFICATION

① 樹幹 TRUNK	② 樹皮 BARK	③ 葉 LEAVES
④ 花 FLOWERS	⑤ 果 FRUITS	

③ 單葉互生，常呈螺旋狀簇生於小枝頂端。葉片革質，倒卵形至寬橢圓形，頂端銳形至短漸尖，基部楔形，葉緣全緣或頂部疏生鋸齒，頂部鋸齒具黑點，中脈凹陷，葉面具光澤，葉背呈淡綠色。葉柄長 5-10 毫米，葉面具凹槽。

Simple leaves alternate, always crowded spirally at the apex of branchlets. Blade leathery, obovate to broadly elliptic, apex acute to shortly acuminate, base cuneate, entire or apically sparsely serrate, black dots at apical teeth, with impressed midvein, adaxially lustrous, abaxially light green. Petioles 5 –10 mm, adaxially grooved.

① 厚皮香的樹幹。

Trunk of *Ternstroemia gymnanthera* (Wight & Arn.) Bedd.

② 樹皮光滑，灰棕色。當年生枝條呈紫紅色至紅棕色，無毛。

Bark smooth, greyish brown. Current year branchlets purplish red to reddish brown, glabrous.

④ 雄花完全花異株。花朵腋生、單生或簇生於無葉枝條。花梗長 1-1.5 厘米。兩性花，花瓣呈淡黃色，雄蕊眾多，花柱較短，頂端具 2 裂。雄花與雌花近似，但雌蕊退化。

Androdioecious. Flowers axillary, solitary or clustered on leafless branchlets. Pedicels 1-1.5 cm. Bisexual flowers: petals pale yellow, stamens many, style short, short, apically 2-lobed. Male flowers: similar, reduced pistil.

⑤ 漿果球形，成熟時轉為紫紅色。

Berries globose, purplish red when mature.

生態 ECOLOGY

　　當一種植物的花只具有雄蕊或雌蕊時，我們便能輕而易舉地判斷其雌雄。當一種植物的花同時具有雄蕊和雌蕊時，我們稱之為雌雄同體。然而，植物的性別只能透過微觀角度上顯明嗎？換言之，我們很難用肉眼分辨出它的性別。當植株的實際性別隱藏在花苞內，我們形容它為「隱性雌雄異株」。厚皮香一直被視為雄花完全花異株（一種同時具有雙性花和雄花的植物），然而雙性狀態仍備受質疑。有部分研究發現，雙性花的花粉粒較雄花的花粉粒大。有趣的是，雙性花的花粉粒沒有萌發孔，並無法形成花粉管，故不能繁殖和結果。雙性花的花粉能得以傳播，不得不提勞苦功高的傳粉者。故此，厚皮香也許是雌雄異株，而不是雄花完全花異株。

We can easily tell if a plant is male or female when its flowers contain stamens or pistils only. We describe a plant as hermaphrodite when its flowers bear both stamens and pistils. However, can the sex expression of a plant demonstrate only on a microscopic level? In other words, we are hard to distinguish its sex expression with bare eyes. Here we describe it as "cryptically dioecious", in which the actual sex is concealed in a small scale. Naked Anther Ternstroemia has been long regarded as andro-dioecious (a plant carrying both bisexual and male flowers); however, the status of bisexuality is questioned. Some studies found that the bisexual flowers carried larger pollen grains than male flowers. Interestingly, the pollen grains of bisexual flowers have no germination pores and cannot form pollen tubes, so they cannot reproduce and bear fruit. The hard-working pollinators have to be mentioned for spreading the pollen of bisexual flowers. Under this scenario, therefore, Naked Anther Ternstroemia may be dioecious instead of andro-dioecious.

黃牛木 又稱：鳥籠木

Yellow Cow Wood | *Cratoxylum cochinchinense* (Lour.) Blume

相片拍攝地點：香港動植物公園、獅子會自然教育中心
Tree Location: Hong Kong Zoological and Botanical Gardens, Lions Nature Education Centre

名字由來 MEANINGS OF NAME

俗名「黃牛木」源自其樹皮顏色神似黃牛。

The common name "Yellow Cow Wood" describes its similar bark colour to the yellow cattle.

應用 APPLICATION

黃牛木的樹皮、根、葉片可入藥，加工成中藥「黃牛茶」，是「廿四味」的重要材料。「黃牛茶」具清熱、利濕、活血化瘀等功效。除藥用功效外，其木材堅硬、耐用、質地細膩；故黃牛木是製作木雕的常用木材。然而，黃牛木生長速度緩慢，為維護當地生物多樣性，應盡可能避免大規模採伐。

黃牛木是中國造林的常用樹種，它對貧瘠土壤具高耐受性，能夠迅速適應和復育貧瘠土地。

本地分佈狀態 DISTRIBUTIONS	原生物種 Native species
原產地 ORIGIN	中國中部和東南部，以及東南亞內陸。 South-Central and Southeast China, Mainland Southeast Asia.
生長習性 GROWING HABIT	落葉灌木或喬木。高度可達 10 米。 Deciduous shrub or tree. Up to 10 m tall.

花果期 月份	1	2	3	4	5	6	7	8	9	10	11	12

花期：本港四月至五月。果期：本港六月。
Flowering period: April to May in Hong Kong. Fruiting period: June in Hong Kong.

The bark, roots and leaves of Yellow Cow Wood are exploited for a traditional Chinese medicine *Hunagniucha*, which is also known as the primary ingredient of "24 flavours". The medicine can relieve heat, dampness, blood stasis and swelling. Apart from the versatile medicinal functions, its wood is hard, durable, textured; therefore, it is always yearned for making wooden sculptures. However, the growth of Yellow Cow Wood is sluggish and rampant logging activities could retard the expansion of its wild population and pose unknown threat to the local biodiversity.

Yellow Cow Wood is one of the framework species of afforestation in China. Its magnificent tolerance to infertile soils allows it to rapidly acclimatize to and revitalize the barren lands.

辨認特徵 TRAITS FOR IDENTIFICATION

① 樹幹 TRUNK	② 樹皮 BARK	③ 葉 LEAVES
④ 花 FLOWERS	⑤ 果 FRUITS	

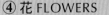

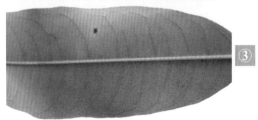

① 黃牛木的樹幹。

Trunk of *Cratoxylum cochinchinense* (Lour.) Blume.

② 樹皮光滑，黃白色至橙棕色，塊狀剝落。

Bark smooth, yellowish white to orange brown, peeling off in lumps.

③ 單葉對生，葉片紙質，橢圓形至長橢圓形或披針形，基部楔形至鈍形，頂端銳形或短漸尖，成熟時由紅色轉為綠色，葉背淡綠色，具腺體或黑點，兩面側脈凸起。

Simple leaves opposite. Blade papery, narrowly elliptic to oblong or lanceolate, base cuneate to obtuse, apex acute or shortly acuminate, abaxially pale green, with glands or black spots near the margin, lateral veins slightly eminent on both surfaces. Young leaves red, green when mature.

91

④ 花腋生或頂生，花朵 2-5 朵，甚少單獨一朵。萼片橢圓形，花瓣 5 片，呈暗緋紅色至粉黃色。雄蕊眾多，通常束成 3 束，呈黃色。

Flowers axillary or terminal, 2– 5 flowered, seldomly 1. Sepals elliptic, Petals 5, crimson. Stamens many, always clustered into 3 distinct bundles, yellow.

⑤ 蒴果橢圓形，被宿存的花萼包被一半以上。成熟時由綠色轉為暗褐色。

Capsules elliptic, more than half of enclosed by persistent calyx. Turning green to dark brown when mature.

生態 ECOLOGY

　　黃牛木的棲息地廣泛，通常分佈在低海拔森林、灌木叢、草地和河岸。黃牛木的葉片和花朵有助豐富當地的生物多樣性，它的葉片是香港常見的蝴蝶品種 —— 萊灰蝶幼蟲的主要食物來源，它的花朵則為蜜蜂和其他本地昆蟲提供豐富花蜜來源。

Yellow Cow Wood is relatively undemanding to environments, showing a wide range of distributions in low-elevation forests, thickets, grasslands and riverbanks. Its leaves and flowers help shape the local biodiversity; its leaves are a pivotal food source to the larvae of *Remelana jangala* (Chocolate Royal), a common butterfly species in Hong Kong; its flowers serve as nectar hotspots for bees and other local animals.

菲島福木

Fukugi, Common Garcinia, Happy Tree | *Garcinia subelliptica* Merr.

相片拍攝地點：金鐘花園、香港動植物公園
Tree Location: Admiralty Garden, Hong Kong Zoological and Botanical Gardens

名字由來 MEANINGS OF NAME

フクギ（Fukugi）是一個日文名稱，意為「幸福樹」，寓意為給人類帶來希望和幸福。為了將這個含意翻譯成中文，故稱為「福木」。除日本外，菲律賓亦是此品種的原產地，故又稱作「菲島福木」。

Fukugi is a Japanese name with a meaning of "Happiness Tree", and is believed to render people prospect; it is directly translated into 「福木」while 「福」refers to fortune in Chinese. Besides Japan, the Philippines is also the origin of this species, therefore it is also named as 「菲島福木」(Filipino Fortune Wood) in Chinese.

本地分佈狀態 DISTRIBUTIONS	外來物種 Exotic species
原產地 ORIGIN	日本、菲律賓、斯里蘭卡和印度尼西亞。 Japan, Philippines, Sri Lanka, and Indonesia.
生長習性 GROWING HABIT	常綠喬木。高度可達 2 米或以上。 Evergreen tree. Up to 2 m or above tall.

1	2	3	4	5	6	7	8	9	10	11	12	花果期 月份

花期：本港三月至八月。果期：本港九月至十一月。
Flowering period: March to August in Hong Kong. Fruiting period: September to November in Hong Kong.

福木對風和鹽分具有中等耐受性，故在日本因不同的目的而被廣泛種植。此品種在沿海地區，例如琉球群島作為防風樹，以減少颱風和海浪造成的損失。同時，日本人相信這樹種可以帶來財富和避邪，故作為風水樹種種植在住宅和神社周邊。值得一提的是，在琉球群島中，有一個接近 300 年歷史的風水林，當中主要種植福木。在沖繩，部分福木因在學術和文化價值上獨樹一幟，而被尊稱作珍稀古樹。

由於福木擁有圓錐形的樹冠和直立的樹幹，這樹種作為觀賞樹被引入至不同國家，在公園和花園中營造春和景明的景色。

它的樹皮可用於製作織物染料。此外，其葉片已被證實具有多種藥效，包括抗癌、抗炎和抗菌。

Fukugi is widely planted in Japan for versatile functions. By virtue of its great resilience to wind and salt, it is always planted in coastal areas such as Ryukyu Islands for attenuating strikes from typhoons and sea waves. The tree is culturally important to Japanese people. With a belief that the tree could bring in wealth and ward off evil spirits, it is regarded as a Feng Shui tree in residential and sacred sites. Notably, there is a Feng Shui woodland in Ryukyu Islands with a stunning 300-year planting history. In Okinawa, some Fukugi are also registered as heritage trees with their exceptional academic and cultural values.

On account of its handsome tree form with symmetrical crown and erect stem, it always serves as an excellent greening component to inject the city with perceived vitality while requires low maintenance cost and less demands to multiple environments.

Its bark can be used to make fabric dyes. In addition, its leaves have been shown to have a variety of medicinal effects, including anti-cancer, anti-inflammatory and antibacterial.

辨認特徵 TRAITS FOR IDENTIFICATION

① 樹幹 TRUNK	② 樹皮 BARK	③ 葉 LEAVES
④ 花 FLOWERS	⑤ 果 FRUITS	

① 菲島福木的樹幹。

Trunk of *Garcinia subelliptica* Merr.

② 樹皮灰色。小枝粗壯而堅硬，具明顯托葉環痕，汁液於破損時流出。

Bark greyish. Branchlets robust and rigid, prominent annular stipular scars, secreting sap when damaged.

③ 單葉對生，葉片卵形、卵狀長橢圓形或橢圓形，厚革質或近肉質，基部寬楔形至近圓形，頂端鈍形，葉緣內捲，中脈於黃綠色葉背凸起，葉面呈深綠色。

Simple leaves opposite. Pseudo-ligule present. Blade thickly leathery, adaxially dark green, ovate, ovate-oblong or elliptic, base broadly cuneate to subrounded, apex obtuse, involute, yellowish green, midvein prominent abaxially.

④ 單生花，雌雄同株。雄花與雌花簇生或單生，花瓣 5 片，呈綠白色。雄花具眾多雄蕊，5 管束生。雌花具長花梗，柱頭盾形。

Monoecious. Male and female flowers clustered or solitary, petals 5, greenish white. Male flowers many stamens, fascicled into 5 bundles. Female flowers with a long pedicel, stigma peltate.

⑤ 漿果近球形，光滑，成熟時由綠色轉為橙黃色。

Berries subglobose, smooth, turning from green to orange yellow at maturity.

生態 ECOLOGY

此品種是哺乳類動物和昆蟲的重要食物來源，同時它亦為鳥類提供棲息地。

This species is an important food source for mammals and insects. It also provides a habitat for birds.

中華杜英 又稱：野杜英

Chinese Elaeocarpus, Elaeocarpus | *Elaeocarpus chinensis* (Gardner & Champ.) Hook. f. ex Benth.

相片拍攝地點：黃泥涌樹木研習徑
Tree Location: Aberdeen Tree Walk

名字由來 MEANINGS OF NAME

杜英屬通稱融合了希臘語 *elaia* （橄欖）和 *karpos*（水果），來統稱其橄欖狀水果。種加詞 *chinensis* 意指華人，形容中華杜英原生於中國。

The generic name *Elaeocarpus* is a blend of Greek words *elaia* (olive) and *karpos* (fruit), collectively describing its olive-like fruits. The epithet *chinensis* means Chinese, and describes the tree was born in China.

應用 APPLICATION

中華杜英具有藥用價值，可用於紓緩月經不調的症狀和治療外傷。

Chinese Elaeocarpus has medicinal value and can mitigate the symptoms of emmeniopathy and traumatic injuries.

本地分佈狀態 DISTRIBUTIONS	原生物種 Native species
原產地 ORIGIN	廣泛分佈在中國中南部和東南部海拔 300-900 米的常綠林中。它亦原生於香港和越南。 Widely distributed in evergreen forests at 300-900 m above sea level in South-Central and Southeast China. Native in Hong Kong and Vietnam.
生長習性 GROWING HABIT	常綠喬木。高度可達 7 米。 Evergreen tree. Up to 7 m tall.

花果期月份	1	2	3	4	5	6	7	8	9	10	11	12

花期：本港五月至六月。果期：本港十月至十一月。
Flowering period: May to June in Hong Kong. Fruiting period: October to November in Hong Kong.

① 樹幹 TRUNK	② 樹皮 BARK	③ 葉 LEAVES
④ 花 FLOWERS	⑤ 果 FRUITS	

① 中華杜英的樹幹。

Trunk of *Elaeocarpus chinensis* (Gardner & Champ.) Hook. f. ex Benth.

② 小枝幼時被微柔毛，成熟時轉為無毛。

Branchlets puberulous at first, glabrescent.

③ 互生，簇生於當年生小枝上。葉片薄革質，幼時被短柔毛，漸變無毛，卵狀披針形或披針形，頂端漸尖，基部圓形，甚少寬楔形，葉緣具微小圓齒，葉背具黑色腺點，側脈 4-6 對，在中脈和側脈交匯處具明顯蟲室。

Simple leaves alternate, crowded at the current branchlets. Blade thin leathery, pubescent at first, glabrescent, ovate-lanceolate or lanceolate, apex acuminate, base rounded, rarely broadly cuneate, minutely crenate, abaxially densely black punctate, lateral veins 4-6 pairs, with observable domatia at the junction of the midvein and the lateral veins.

④ 雜性植物（雄花和雙性花）。總狀花序，腋生。兩性花：萼片 4 片，花瓣 4 片，呈綠白色，雄蕊 8-10 條。雄花：萼片和花瓣與兩性花一致，雄蕊 8-10，雌蕊退化。

Polygamous (male and bisexual flowers). Racemes axillary. Hermaphrodite flowers: sepals 4, petals 4, greenish white, stamens 8-10. Male flowers: sepals and petals as bisexual flowers, stamens 8-10, without pistillodes.

⑤ 核果橢球形，成熟時轉為黑色。

Drupes ellipsoid, turning black at maturity.

植物趣聞 ANECDOTE ON PLANTS

蟲室 Domatia：

　　與動物不同，植物是靜態的，對外來攻擊十分脆弱。在物競天擇，適者生存的情況下，植物發展出多種防禦系統以應付天敵。蟲室是位於中脈脈腋和葉背側脈的改良結構，使植物能安全地「僱用」蟎蟲為其「貼身保鏢」，讓其潛伏於蟲室之中，悄悄吞噬出現在葉子上的害蟲（例如薊馬和粉蝨等）。

　　在不同的物種中，蟲室會武裝成不同的形態。例如欖仁樹和秋楓的蟲室呈坑狀。大多數杜英屬具有三角形口袋狀的蟲室，例如中華杜英、山杜英、杜英及日本杜英。

　　顯微世界的植物趣味橫生，亦可從中得到啟發。現在不妨拿起你的放大鏡，一探這些大自然奇妙的奧祕。

Unlike animals, plants are sessile and susceptible to attacks. Under the intensive and endless selective pressure, plants have developed versatile defence systems to tackle their natural enemies. Domatia are types of modified leaf structures at the junction of midveins and lateral veins of abaxial leaves, with a purpose of maintaining the plant safe by hiring "bodyguards" which are usually mites. The bodyguards conceal themselves under the domatia and quietly devour pests (e.g. thrips and white flies) that appear on the leaves.

Domatia of plants are with multiple shapes. For example, the domatia of *Terminalia catappa* (Indian Almond) and *Bischofia javanica* (Autumn Maple) are pit-like. Not all but most of *Elaeocarpus* are equipped triangular-pocket-like domatia, such as Chinese Elaeocarpus, *E. sylvestris* (Woodland Elaeocarpus), *E. decipiens* (Common Elaeocarpus) and *E. japonicus* (Japanese Elaeocarpus).

The microscopic world is always amazing and inspiring. It is time to pick up your magnifying glass and explore the amazing secret of the nature.

山杜英 又稱：膽八樹、羊屎樹、羊仔樹

Woodland Elaeocarpus | *Elaeocarpus sylvestris* (Lour.) Poir.

相片拍攝地點：獅子會自然教育中心
Tree Location: Lions Nature Education Centre

名字由來 MEANINGS OF NAME

　　杜英屬植物的通稱是希臘語 *elaia*（橄欖）和 *karpos*（水果）的合稱，用來描述它類似橄欖的水果。種加詞 *sylvestris* 具樹木繁茂或生長在荒野的含意，暗指山杜英的生長特性。

The generic name *Elaeocarpus* is a blend of Greek words *elaia* (olive) and *karpos* (fruit), describing its olive-like fruits. The specific epithet *sylvestris* means wooded or growing in the wild.

應用 APPLICATION

　　山杜英一直以來因其藥用特性而備受青睞。據《本草綱目》記載，山杜英的種子油可加工成祭祀用香，用以驅邪辟邪，亦可加工成肥皂和潤滑油。其根部可有效紓緩跌倒和碰撞引起的腫脹。其葉片因具抗氧化功效而備受重視。

本地分佈狀態 DISTRIBUTIONS	原生物種 Native species
原產地 ORIGIN	中國廣東、海南、廣西、福建、浙江，江西、湖南、貴州、四川及雲南等省份，越南亦有分佈。 The provinces of Guangdong, Hainan, Guangxi, Fujian, Zhejiang, Jiangxi, Hunan, Guizhou, Sichuan and Yunnan in China, and also Vietnam.
生長習性 GROWING HABIT	常綠喬木。高度可達 15 米。 Evergreen tree. Up to 15 m tall.

1	2	3	4	5	6	7	8	9	10	11	12	花果期 月份

花期：本港四月至五月。果期：本港五月至八月。
Flowering period: April to May in Hong Kong. Fruiting period: May to August in Hong Kong.

山杜英的葉片顏色於年老時會由綠色轉為紅色。作為澳門和台灣常見的行道樹之一，它為整個城市渲染了天然亮麗的紅色，令人目不暇給。

Woodland Elaeocarpus has been long appreciated for its excellent medicinal properties. According to the *Compendium of Materia Medica*, the extracted seed oil can be processed into incense to expel evil spirits. The oil can be also used for making soap and lubricant. The roots are medicinally effective to relieve swelling from falls and knocks.

Its leaves turn from green to red as the tree ages. As one common tree species planted along the pedestrian ways in Macao and Taiwan, it renders the whole city a natural and bright red color, which is dizzying for the eyes.

辨認特徵 TRAITS FOR IDENTIFICATION

① 樹幹 TRUNK　　② 樹皮 BARK　　③ 葉 LEAVES
④ 花 FLOWERS

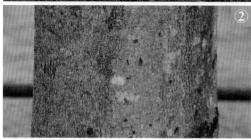

① 山杜英的樹幹。

Trunk of *Elaeocarpus sylvestris* (Lour.) Poir.

② 樹皮褐色。小枝纖細，被微柔毛或疏柔毛，變乾時呈黑褐色。

Bark brownish. Branchlets slender, puberulent or sparsely pilose, blackish brown when dry.

③ 單葉互生。葉片紙質，無毛，倒卵形或倒披針形，基部狹楔形、下延，頂端鈍形或短漸尖，葉緣具鈍鋸齒。側脈 4-5 對，於葉背突出，在中脈和側脈的連接處具明顯蟲室。

Simple leaves alternate. Blade papery, glabrous, obovate or oblanceolate, base narrowly cuneate and decurrent, apex obtuse or shortly acuminate, margin crenate. Lateral veins 4-5 pairs, prominent abaxially, with observable domatia at the junction of the midvein and the lateral veins.

④ 雌雄同體。總狀花序，腋生，下垂。萼片5片，披針形，花瓣5片，每瓣具10條裂片，呈白色。雄蕊花藥無毛叢或附屬物，15條。

Hermaphroditic. Racemes axillary, drooping. Sepals 5, lanceolate, petals 5, each with 10 segments, white. Stamens 15, without awns at apex.

註：本樹另有核果，橢圓形，細小，成熟時轉為黑色。

Remarks: Drupes ellipsoid, small, black when mature.

生命力 VITALITY

　　山杜英廣泛分佈於海拔300-2000米的常綠森林。其生長迅速，能在不同環境中茁壯成長，但偏愛溫暖潮濕的氣候。

Woodland Elaeocarpus is widely distributed in evergreen forests at 300-2000 m above sea level. It is fast-growing and can acclimatise to multiple environments, while good planting prefers warm and moist weather.

植物趣聞 ANECDOTE ON PLANTS

杜英屬的迷思 Confusion in *Elaeocarpus*：

　　由於山杜英、杜英和禿瓣杜英的特徵極為相似，難以辨別，故常被混為一談。事實上，許多植物分類學家亦曾將三者視為同一物種，將杜英和禿瓣杜英視為山杜英的別名。

　　在最新的分類學修訂中，這三種因形態差異而已被肯定各有不同。禿瓣杜英的小枝和花瓣無毛，而杜英和山杜英的小枝和花瓣則被微柔毛。為了區分杜英和山杜英，可從側脈和雄蕊數量入手。杜英具7-9對側脈和25-30條雄蕊，山杜英則只有4-5對側脈和15條雄蕊。未來更進一步的系統發育研究將會對這三種杜英的分類提供更深入和確實的證明。

Woodland Elaeocarpus is easily confused with E. decipiens (Common Elaeocarpus) and E. glabripetalus with their imperceptibility in traits. Many plant taxonomists tend to lump them in the same species, considering Common Elaeocarpus and E. glabripetalus as the synonyms of Woodland Elaeocarpus.

In the updated taxonomic revision, however, they were suggested being split but not lumped together, giving by their unmistakably diagnosed keys. E. glabripetalus stands out from showing glabrous branchlets and petals, while those of Common Elaeocarpus and Woodland Elaeocarpus are rather puberulent. To distinguish the latter two Elaeocarpus, Common Elaeocarpus has 7-9 lateral veins and 25-30 stamens, while Woodland Elaeocarpus has only 4-5 lateral veins and around 15 stamens. A further phylogenetic study would be an advanced and authentic testament to the taxonomical statues of the three Elaeocarpus.

破布葉 又稱：布渣葉

Microcos, Panicilate Microcos | *Microcos nervosa* (Lour.) S. Y. Hu

相片拍攝地點：獅子會自然教育中心、香港墳場
Tree Location: Lions Nature Education Centre, Hong Kong Cemetery

名字由來 MEANINGS OF NAME

　　破布葉的葉片質地似縐布，其屬名 *Microcos* 即為「縐布」之意。

The generic name *Microcos* refers to its leaf texture akin to crepe cloth.

應用 APPLICATION

　　破布葉嫩枝上的葉片經採摘曬乾後可入藥。其葉片具豐富類黃酮，如：異鼠李素、山奈酚和槲皮素。味酸、性平，可緩解感冒、腹瀉和腹腔黏連等病徵。就其藥用價值而言，破布葉是製作「廿四味」的藥材之一。除醫藥用途外，破布葉的木材還可加工成建築用料，而樹皮則可用於製作繩索。

本地分佈狀態 DISTRIBUTIONS	原生物種 Native species
原產地 ORIGIN	中國華南省份、巴基斯坦、馬來西亞等地區。 The provinces of south China, Pakistan and Malaysia.
生長習性 GROWING HABIT	常綠灌木或小喬木。高度可達 12 米。 Shrub or small evergreen tree. Up to 12 m tall.

花果期 月份	1	2	3	4	5	6	7	8	9	10	11	12

花期：本港六月至七月。果期：本港七月至十二月。
Flowering period: June to July in Hong Kong. Fruiting period: July to December in Hong Kong.

The leaves are usually dried for traditional Chinese medicines. They contain abundant flavonoids, such as isorhamnetin, kaempferol and quercetin; therefore, they taste sour, tasteless and with neutral medicinal properties. The leaves can effectively relieve common cold, diarrhea and abdominal adhesion. As a result of its excellent medicinal effects, Microcos is also one of the primary ingredients of "24 flavours", which is a well-known Cantonese herbal tea. Other than medicinal functions, the wood of Microcos can be processed into building materials while the bark can be used for making ropes.

辨認特徵 TRAITS FOR IDENTIFICATION

① 樹幹 TRUNK	② 樹皮 BARK	③ 葉 LEAVES
④ 花 FLOWERS	⑤ 果 FRUITS	

① 破布葉的樹幹。

Trunk of *Microcos nervosa* (Lour.) S.Y. Hu.

② 樹皮粗糙，呈黃棕色或灰棕色，小枝披絨毛。

Bark coarse, yellowish brown or greyish brown , branchlets hairy.

③ 葉片互生，薄革質至紙質，卵形或卵狀長橢圓形，基部圓形，頂端漸尖，葉緣呈不規則細鋸齒狀或波狀，葉脈明顯，三出脈。托葉宿存，線狀披針形，葉柄被柔毛，腫脹。

Simple leaves alternate. Blade thin leathery to papery, ovate or ovate oblong, base round, apex acuminate, irregularly tiny-toothed or undulate, stellate pubescent initially, glabrescent, lateral veins obvious, trinerved. Stipules persistent, linear-lanceolate.

④ 圓錐花序，頂生或腋生。雙性花，花瓣 5 片，被黃色短柔毛。

Hermaphroditic. Panicles terminal or axillary, stellate pubescent. Petals 5, yellowish pubescent.

⑤ 核果，近球形或倒卵圓形，無毛，成熟時由綠色轉為黑棕色。

Drupes subglobose or obovoid, glabrous, turning dark brown when ripe.

生態 ECOLOGY

　　破布葉是一種已在香港常見的歸化品種，廣泛分佈於灌木叢和山坡上。此物種與本地野生動物有密不可分的關係。例如，它的葉片為一種栗色的小型蝴蝶 —— 角翅弄蝶的幼蟲提供天然庇護所。角翅弄蝶通常在葉面產卵。幼蟲孵化後會以葉片為食，並能夠捲起葉片以躲避天敵。當幼蟲成長化蛹時，葉片構成的庇護所會脹起。在蛹中掙扎幾天後，美麗的蝴蝶便會破蛹而出。角翅弄蝶的生命週期展示了其與破布葉之間的共生關係。這種生物間的互相作用，破布葉對角翅弄蝶的繁殖成長帶來好處，但本身卻無任何益損。

Microcos is a common tree distributed along thickets and hillslopes in Hong Kong. The tree plays a vital role in environments. For instance, its leaves provide natural shelter to the larvae of *Odontoptilum angulatum* (Chestnut Angle), a small butterfly species with chestnut colour. Chestnut Angle always lay eggs on the leaf surface. After the larvae were born, they feed on the leaves and twist the leaves into a shelter that can conceal themselves from natural enemies. The leaf shelters are swollen when the larvae grow, they ultimately transform into pupae inside the shelters. After days of struggling in a pupa, a beautiful butterfly emerges. The life cycle of Chestnut Angle is an excellent example delineating the commensalism relationship with Microcos, a kind of biological interactions depicting a species benefiting from another species, while there is no any gain or loss to the latter.

槭葉蘋婆

Flame Bottletree | *Brachychiton acerifolius* (A. Cunn.) Macarthur & C. Moore

相片拍攝地點：添馬公園
Tree Location: Tamar Park

名字由來 MEANINGS OF NAME

屬名 *Brachychiton* 意指短外衣，暗指種子上的剛毛。槭葉蘋婆的葉形與楓葉相似，故被冠名為 *acerifolius*（中文意譯為楓葉），亦有「槭葉」此中文名稱。

The generic name *Brachychiton* means short tunic, alluding to its overlapping bristles on seeds. The specific epithet *acerifolius* and the Chinese common name 「槭葉」refers to its maple-like leaves.

應用 APPLICATION

當地人通常收集槭葉蘋婆的種子，並烘乾以作食用。從內層樹皮中提取的纖維可用於製作漁網和魚線。槭葉蘋婆的葉提取物可用作抗氧化劑、香水和護膚化妝品。

本地分佈狀態 DISTRIBUTIONS	外來物種 Exotic species
原產地 ORIGIN	澳洲的新南威爾士州和昆士蘭州。 New South Wales and Queensland in Australia.
生長習性 GROWING HABIT	落葉喬木。高度可達 15 米（野外可達 35 米）。 Deciduous tree. Up to 15 m tall (35 m in the wild).

1	2	3	4	5	6	7	8	9	10	11	12

花果期月份

花期：本港四月。果期：不詳。
Flowering period: April in Hong Kong. Fruiting period: Unknown.

雖然槭葉蘋婆的中文名稱與「假蘋婆」與「蘋婆」相近，在演化學上，它們毫無瓜葛。與假蘋婆和蘋婆相比，槭葉蘋婆那豔紅的花朵為城市增添大自然的色彩。在香港，槭葉蘋婆的花期格外漫長，可長達半年，故此它們被視為觀賞樹木廣泛種植在大街小巷中。在花期時，一整排的槭葉蘋婆掛滿嬌艷欲滴的紅色花朵，形成一片春色，春和景明。有趣的是，若果去年冬天的降雨稀少，花朵便會愈發艷紅如火。

The seeds are esculent and always roasted for food in its native range. Fibre extracted from the inner bark can be used for making fishing nets and fishing lines. The leaf extracts are the ingredients of antioxidants, perfumes and skin conditioners.

Although its Chinese name is similar to 「假蘋婆」 (*Sterculia lanceolata*; Scarlet Sterculia) and 「蘋婆」 (*Sterculia nobilis*; Common Sterculia), they are phylogenetically distinct from each other. Compared with the flowers of Scarlet Sterculia and Common Sterculia, those of Flame Bottle Tree are highlighted in appealing scarlet flowers. In Hong Kong, Flame Bottle Tree is prolific, demonstrating an astonishingly long period of floral display, blooming for almost half of a year. During the flowering seasons, the flamboyant scarlet flowers profusely embellish the crown and carpet the road when dropped. Interestingly, the ornamental effect can be bolstered when the rainfall of last winter was limited.

辨認特徵 TRAITS FOR IDENTIFICATION

① 樹幹 TRUNK	② 樹皮 BARK	③ 葉 LEAVES
④ 花 FLOWERS	⑤ 果 FRUITS	

① 槭葉蘋婆的樹幹。

Trunk of *Brachychiton acerifolius* (A. Cunn.) Macarthur & C. Moore.

② 樹皮呈灰色，光滑，具明顯葉痕。小枝呈綠色。

Bark grey, smooth, with observable leaf scars. Branchlets green.

③ 單葉互生，葉片形狀多變，成熟時葉片常為卵形，葉緣全緣，或呈掌狀 3-7 裂。

Simple leaves alternate. Blades glabrous, variable in shapes, mature trees always ovate and entire, while juvenile trees always palmate with 3-7 lobes.

④ 雌雄同株。圓錐花序頂生於無葉小枝上。花朵缺少花瓣，只有 5 片部分合生的萼片，鐘形，呈朱紅色。

Monecious. Panicles terminal on leafless branchlets. Flowers without petals, sepals 5, partially fused, campanulate, scarlet.

⑤ 木質蒴果，成熟時轉為暗棕色或深色。種子棕黃色，被帶刺剛毛。

Capsules woody, dark brown or dark when ripe.Seeds brownish yellow, covered with stinging bristles.

生命力 VITALITY

　　檄葉蘋婆偏好亞熱帶氣候、乾旱及全日照環境。根部容易腐爛，極易受到澇漬土壤的影響。

Flame Bottletree prefers subtropical climate, with drought and full-sun weather. The roots are easily rotted and extremely susceptible to waterlogged soils.

銀葉樹

Coastal Heritiera | *Heritiera littoralis* Aiton

相片拍攝地點：迪欣湖活動中心
Tree Location: Inspiration Lake Recreation Centre

名字由來 MEANINGS OF NAME

　　屬名 *Heritiera* 是為了紀念法國植物學家夏爾·路易·萊里捷·德布呂泰勒（1746-1800）；種加詞 *littoralis* 顯示它的生長環境為海岸；而中文名稱「銀葉樹」意指其葉背獨有的銀白色細小鱗片。

本地分佈狀態 DISTRIBUTIONS	原生物種 Native species
原產地 ORIGIN	野生銀葉樹分佈在香港大埔滘、榕樹澳和荔枝窩；亦廣泛分佈於亞洲東南部、東非及澳洲。 In Hong Kong, the wild Coastal Heritiera is found in Tai Po Kau, Yung Shue O and Lai Chi Wo. It is also widely distributed throughout Southeast Asia, east Africa and Australia.
生長習性 GROWING HABIT	常綠喬木。高度可達 10 米。 Evergreen tree. Up to 10 m tall.

花果期
月份

1	2	3	4	5	6	7	8	9	10	11	12

花期：本港四月至五月。果期：本港八月至三月。
Flowering period: April to May in Hong Kong. Fruiting period: August to March in Hong Kong.

The genus name *Heritiera* is in honor of the French botanist, Charles Louis L'Héritier de Brutelle (1746-1800), whereas the species name *littoralis* indicates its growing habitat on seashores. The Chinese name「銀葉樹」refers to its abaxial leaves with silver-white scurfy scales.

應用 APPLICATION

銀葉樹強壯結實的木材能用作製造船隻。它亦是海濱公園常見的觀賞性樹種之一。

Its wood is strong and can be used for ship making. In addition, it is commonly planted in waterfront parks as an ornamental tree.

辨認特徵 TRAITS FOR IDENTIFICATION

① 樹幹 TRUNK	② 樹皮 BARK	③ 葉 LEAVES
④ 花 FLOWERS	⑤ 果 FRUITS	

① 銀葉樹的樹幹。具強壯的板根。

Trunk of *Heritiera littoralis* Aiton. Roots butressed, strong.

② 樹皮粗糙，呈灰褐色。小枝幼株時被銀白色的細小鱗片。

Bark gray-brown, rough. Branchlets covered by white scurfy scaly when young.

③ 單葉互生，革質，葉片呈矩圓狀披針形、橢圓形或卵形，頂端急尖或鈍形，基部鈍，全緣。葉面無毛，葉背密被銀白色鱗片。

Leaves alternate, leathery, blades oblong-lanceolate, elliptic or ovate, apex acute or obtuse, base obtuse, margin entire, adaxially glabrous, abaxially densely silver-white scurfy scaly.

④ 圓錐花序，花呈紅棕色，細小而數量多，密被銀白色毛。雌雄同株，單性花。

Panicles, flowers small, many, red-brown, covered by densely silver-white hairs. Monecious, unisexual flowers.

⑤ 蒴果，堅果狀，木質，乾時呈黃褐色，背部隆起，形似卡通人物「超人奧特曼」（又稱：鹹蛋超人）。種子卵球形。

Capsule, nut-like, woody, yellow-brown when dried, ridge on back, like the cartoon character "Ultraman".

生態 ECOLOGY

銀葉樹已演化出一套適合於海岸惡劣環境蓬勃生長的能力。首先，其堅固強壯的板根有助抓緊沙質土壤。此外，葉背銀白色的細小鱗片能夠反射海面的陽光，使葉片保持最佳的溫度和濕度。其果實亦可漂浮在海上，隨波逐流，遠播他鄉。然而，伴着海洋資源的過度開發、全球暖化和水質污染等日趨嚴重，野生銀葉樹的數量也明顯減少。它在國際自然保護聯盟（IUCN）瀕危物種紅色名錄中更是被列為「低關注度物種」。除了銀葉樹外，大量棲息在紅樹林的生物也面臨棲息地減少等的威脅。物種多樣性對於刻畫出多姿多采的自然美景，和豐富本地的生物多樣性至關重要。為了保護這怡人的景色，我們應該收起對自然資源無窮無盡的貪念了。

Coastal Heritiera has developed the ability to thrive on the coast. First, its robust buttresses help to anchor the tree on scales soils. In addition, its abaxial silver-white scurfy scales can reflect the sun glitter from the sea surface and maintain the leaves in optimal temperature and moisture. Its seeds can travel for a long distance by floating on sea and driven by sea waves. However, with the overexploitation of marine resources, global warming and water pollution, the population of Coastal Heritiera is reduced perceptibly. It is currently registered as a threatened species on the IUCN Red List in status of "Least Concern". Indeed, not only Coastal Heritiera, species in bulk growing in the coastal intertidal zone are under great threat from the impacts of e.g. the loss of habitats. Species diversity is important to imprint an exquisite natural scenery and enrich the local biodiversity. To preserve this beautiful scene, it is time to control our insatiable desire for natural resources.

生命力 VITALITY

此物種對高鹽度和澇漬土壤具有良好的耐受性。

This species has good tolerance to high salinity and waterlogged soils.

梭羅樹 又稱：兩廣梭羅

Reevesia, Buch-like Reevesia | *Reevesia thyrsoidea* Lindl.

相片拍攝地點：九龍公園、獅子會自然教育中心、香港動植物公園、荔枝角公園
Tree Location: Kowloon Park, Lions Nature Education Centre, Hong Kong Zoological and Botanical Gardens, Lai Chi Kok Park

名字由來 MEANINGS OF NAME

屬名 *Reevesia* 是紀念首次發現此品種的英國自然學家 —— 約翰·里夫斯（1774-1856）。

The generic name *Reevesia* is named after John Reeves (1774-1856), an English naturalist who made the first discovery of the tree.

應用 APPLICATION

由於梭羅樹的花朵香氣撲鼻、具艷麗花色，除了賞心悅目，也是昆蟲的重要食物來源，使其被廣泛栽培在城市公園和花園中。此外，梭羅樹的木材堅硬，適合用於製作家具；樹皮結實，可加工成繩索、麻袋和造紙材料。

本地分佈狀態 DISTRIBUTIONS	原生物種 Native species
原產地 ORIGIN	原生於中國南部，西南至中部省份，越南和柬埔寨也有分佈。 The provinces of South, Southeast and Central China, Vietnam and Cambodia.
生長習性 GROWING HABIT	常綠喬木。高度可達 20 米。 Evergreen tree. Up to 20 m tall.

1	2	3	4	5	6	7	8	9	10	11	12

花果期月份

花期：本港三月至四月。果期：本港六月至十月。
Flowering period: March to April in Hong Kong. Fruiting period: June to October in Hong Kong.

The flowers of Reevesia are fragrant with glamorous flowers. Besides looking pleasing to the eyes, the flowers also serve as an important food source to many insects. The tree is therefore widely planted in urban parks and gardens. The wood is hard and suitable for making furniture. The bark is strong and can be processed into ropes, gunny bags and paper-making materials.

辨認特徵 TRAITS FOR IDENTIFICATION

① 樹幹 TRUNK	② 樹皮 BARK	③ 葉 LEAVES
④ 花 FLOWERS	⑤ 果 FRUITS	

① 梭羅樹的樹幹。

Trunk of *Reevesia thyrsoidea* Lindl.

② 樹皮塊狀，粗糙，呈灰棕色，通常具菱形裂紋。

Bark grey-brown, coarse, lumpy, always cracked in rhombuses.

③ 葉片互生，聚生於枝條頂端。葉片披針形或卵狀披針形，頂端銳形或漸尖，革質，葉面呈翡翠綠，具光澤，葉背呈淺綠色。葉柄兩端腫脹。

Simple leaves alternate, aggregated at the apex of branchlets. Leaf blade oblong to elliptic, apex acute or acuminate, adaxially emerald and glossy, abaxially pastel green. Petioles swollen at two proximal ends of petioles.

④ 傘房花序，頂生，花朵密集。花瓣 5 片，呈白色。雌雄蕊柄頂端帶 15 條花藥。

Corymbose, terminal, densely flowered. Petals 5, white. Androgynophore with 15 anthers at apex.

⑤ 蒴果，橢球形，具 5 棱，被短柔毛。種子密集，具翅。

Capsules ellipsoid, 5-ribbed, pubescent. Seeds compressed and winged.

植物趣聞 ANECDOTE ON PLANTS

梭羅樹和假蘋婆 Reevesia and *Sterculia lanceolata* (Lance-leaved Sterculia)：

梭羅樹和假蘋婆皆屬於梧桐科，兩者特質相似，難以分辨。要清楚分辨兩者，可從樹幹及葉背入手。梭羅樹的葉背呈淡綠色，而假蘋婆的葉背則呈綠色，並帶有明顯網狀脈。梭羅樹的樹皮裂成菱形，而假蘋婆的樹皮則沒有這種圖案。

Two species are the members of Sterculiaceae and are always confused with their comparable characteristics without blossoms. To distinguish the trees, we can observe the trunk and abaxial leaves. The abaxial leaves of Reevesia are pastel green, while Lance-leaved Sterculia are green with discernible netted veins. The bark of Reevesia is cracked in rhombuses, while this pattern is absent in Lance-leaved Sterculia.

吉貝 又稱：美洲木棉

Kapok Ceiba, Silk Cotton Tree | *Ceiba pentandra* (L.) Gaertn.

相片拍攝地點：維多利亞公園
Tree Location: Victoria Park

應用 APPLICATION

　　吉貝的木材質輕柔韌，可加工成火柴和木屐。含絲狀棉毛的果實具強勁浮力和低吸水性，可用於製作枕頭、救生衣和床墊的填充物料。同時亦可提煉種子油來製造肥皂和生物柴油。吉貝有令人眼前一亮的樹形和闊大樹冠，是一種具有遮蔭效果的主要行道樹種。

The light and flexible wood is always processed into matches and clogs. The fruits yielding profuse silky wool can be used for fillers of pillows, life jackets and mattresses due to their enormous buoyancy and low water-absorption. The seeds are extracted for making soap and producing biodiesel. With a striking tree shape and broad crown, it is a main street tree species that provides shading effect.

本地分佈狀態 DISTRIBUTIONS	外來物種 Exotic species
原產地 ORIGIN	熱帶美洲，如巴哈馬、墨西哥和尼加拉瓜。 Tropical America, such as Bahamas, Mexico and Nicaragua.
生長習性 GROWING HABIT	落葉喬木。高度可達 30 米或以上。 Deciduous tree. Up to 30 m or above tall.

花果期 月份

1	2	3	4	5	6	7	8	9	10	11	12

花期：本港三月至四月。果期：本港五月至七月。
Flowering period: March to April in Hong Kong. Fruiting period: May to July in Hong Kong.

① 樹幹 TRUNK	② 樹皮 BARK	③ 葉 LEAVES
④ 花 FLOWERS		

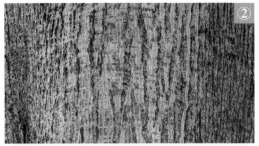

① 吉貝的樹幹。老樹具強壯板根。

Trunk of *Ceiba pentandra* (L.) Gaertn. Strong buttresses root in old trees.

② 樹幹初時具刺，樹皮灰棕色。小枝輪生，水平開展。

Bark greyish brown, stem spiny when young. Branchlets verticillate, spreading horizontally.

③ 掌狀複葉互生，具 5-9 片小葉。小葉薄革質，無毛，長橢圓狀披針形，頂端銳形或漸尖，基部楔形，葉緣全緣或具微鋸齒。

Palmately compound alternate, leaflets 5-9. Blade thin-leathery, glabrous, oblong-lanceolate, apex acute or acuminate, base cuneate, entire or remotely and minutely toothed toward the apex.

④ 花先葉或與葉同時開放，多數簇生於上部葉腋間。花瓣 5 片，呈乳白色。花朵常於夜間盛放，翌日下午時分凋落。

Flowers bloom before leaves emerge or together with leaves. Flowers fascicled in leaf axils on upper part of twigs, subterminal, solitary or in fascicles. Petals 5, cream. Blossoming at nighttime and shedding or closed before next afternoon.

註：本樹另有大型蒴果，橢圓形，向上漸狹，密被絲狀棉毛，5 瓣開裂，成熟時由綠色轉為棕色。種子呈黑色，球狀。

Remarks: Capsules large, ellipsoid, tapering toward tip, filled by dense silky wool, brown and dehiscing into 5 valves at maturity. Seeds black, globose.

類似於 *Kigelia africana*（吊燈樹），吉貝也是一種夜間開花的夜行性植物。晚上，它們繁花錦簇，同時會分泌高濃度的花蜜，以配合夜行性動物的活動習性。蝙蝠（例如犬蝠）是這品種的主要傳粉者。蝙蝠會在飛行過程中覓食花蜜，並傳播花粉。花朵會在白天閉合或掉落，晝行性動物只能採食黏附在花朵邊緣的花蜜，對授粉效能較小。

Like *Kigelia africana* (Sausage Tree), Kapok Ceiba is also a nocturnal tree. At nighttime, the tree blossoms and the flowers release high concentration of nectar to match with the habits of nocturnal animals. Bats (e.g. Greater Short-nosed Fruit Bat) are the main pollinators of this species. They forage nectar during their flight and pollinate. The flowers are closed or detached at daytime, and therefore the diurnal animals can only feed on the nectar around the edge of the flowers, which is less effective for pollination.

國樹 National tree：

吉貝為危地馬拉和波多黎各的國樹。

Kapok Ceiba is the national tree of Guatemala and Puerto Rico.

神聖的名木 Sacred and valuable tree：

吉貝是一種快速生長的樹種，最高可達 30 米，樹幹直徑為 3 米。廣闊的根系和板根有助於支撐樹木並吸取維持生長的養分。通常野生吉貝在樹冠層中尤為突出。由於此品種具標誌性的高度，故被古瑪雅人推崇為世界中心的聖樹，掌控宇宙，接天連地。

Kapok Ceiba is fast-growing and can attain a stunning 30 m. The extensive root system and stiff buttress give support to the trees and promote the absorption of nutrients for growing. Wild Kapok Ceiba is always protruded from the upper canopy. Considering its majestic character, the tree was worshipped by the ancient Maya people who believed that the tree was grown in the centre of the world, grasped the universe and built a connection from world to heaven.

在香港，沿着維多利亞公園周邊種植的吉貝因其驚人的樹高和胸徑而被登記為古樹名木（OVT）。其中一棵（編號：LCSD WCH/29）為目前香港已知最大的吉貝，其胸徑為 1360 毫米，高度為 33.5 米，樹冠為 20 米。

In Hong Kong, a row of Kapok Ceiba planting along the periphery of Victoria Park render an aura of wonderful shading effect. The trees are registered as Old and Valuable Trees (OVT) in respect of their astonishing tree height and diameter at breast height. One of them (LCSD WCH/29) is currently the most gigantic Kapok Ceiba in Hong Kong, with DBH of 1360 mm, height of 33.5 m and crown spread of 20 m.

吉貝古樹名木

絲木棉 又稱：美人樹、美麗異木棉
Silk Floss Tree, Floss Silk Tree | *Chorisia speciosa* A. St.-Hil.

相片拍攝地點：香港動植物公園
Tree Location: Hong Kong Zoological and Botanical Gardens

名字由來 MEANINGS OF NAME

種加詞 *speciosa* 有華麗的意思，意指其秀麗的花朵。絲木棉的種子被一層絲質纖維包裹着，故名為「絲木棉」。

The specific epithet *speciosa* means "gorgeous" and refers to its beautiful blossoms. The common name "Silk Floss Tree" describes its seeds are wrapped by a layer of silky fibres.

應用 APPLICATION

絲木棉是一種令人眼前一亮的觀賞樹種，其花色艷麗，可印上一道道絕美的山水畫卷。此外，它的綿花還可用作枕頭、靠墊和背心的填充物。

Silk Floss Tree is a superb ornamental tree with its spectacular blossoms. In addition, its floss is nice to be fillers of pillows, cushions and vests.

本地分佈狀態 DISTRIBUTIONS	外來物種 Exotic species
原產地 ORIGIN	南美洲。 South America.
生長習性 GROWING HABIT	落葉喬木。高度可達 15 米。 Deciduous tree. Up to 15 m tall.

1	2	3	4	5	6	7	8	9	10	11	12

花果期月份

花期：本港十月至二月。果期：本港十二月至二月。
Flowering period: October to February in Hong Kong. Fruiting period: December to February in Hong Kong.

① 樹幹 TRUNK	② 樹皮 BARK	③ 葉 LEAVES
④ 花 FLOWERS	⑤ 果 FRUITS	

① 絲木棉的樹幹。

Trunk of *Chorisia speciosa* A. St.-Hil.

② 樹幹具短刺，基部膨脹。樹皮光滑，成熟時由綠色轉成灰白色。

Trunk with short spines, bulging basal stem. Bark turns from green to off-white as it matures.

③ 掌狀複葉，互生，螺旋排列於小枝上。小葉 6-7 片，紙質，葉片長圓狀倒披針形，葉緣具鋸齒，中間葉片較大。

Leaves palmately compound alternate, spirally arranged on the current branchlets, leaflets 6-7, papery, leaf blade oblong and needle-shaped, margin serrate, middle blade always larger.

④ 單生花或 2-3 朵簇生於頂端葉腋。花瓣 5 片，呈粉紅色，微黃色，正面白色具深色條紋。

Solitary or 2-3 crowded at axils of apical leaves. Petals 5, pink, slightly yellow, adaxially white with dark stripes.

⑤ 蒴果橢圓形。種子密被絲綢棉毛。

Capsules ellipsoid. Seeds covered with dense silky wool.

絲木棉獨有的生長特徵能作為優秀的生態教學材料。經進化後，絲木棉的樹幹發展出帶有圓錐形的刺，能夠避免動物攀上樹冠咬爛葉子。此外，絲木棉凸起的樹幹能儲存水分，以適應原生乾燥的環境。其嬌艷欲滴的花朵能吸引蝴蝶幫忙授粉。當果實成熟時，被絲絨包裹的種子透過風傳播至他方。

Silk Floss Tree is a proper teaching material for studying ecology. First, the trunk is covered with conical prickles which are modified stems for the sake of expelling animals from climbing onto the tree and biting its leaves. In addition, its bulging stem is meticulously designed for water storage and allows the tree to acclimatise to drought seasons in its origin. The flowers are appealing to many pollinators. When the fruits are ripe, the seeds enclosed in silk floss can catch breeze for journeys of miles.

絲木棉生長迅速，對鹽鹼、乾旱和風具有高耐受性。

Silk Floss Tree grows rapidly and its highly resistant to salinity, drought and wind.

絲木棉，吉貝和木棉 Silk Floss Tree, *Ceiba pentandra* (Kapok Ceiba) and *Bombax ceiba* (Tree Cotton)：

木棉、吉貝和絲木棉皆屬於木棉科，它們是香港市區常見樹種，但如果缺少深入觀察，便會很容易將三者混淆。要清楚分辨三者，可從樹幹、葉片、花朵及果實入手。

雖然三者的樹幹皆帶有圓錐狀的刺，但吉貝上的刺更為明顯，而絲木棉的樹幹基部更為腫脹。在三者當中，木棉的葉最大而且葉緣全緣；吉貝葉的大小僅次於木棉，並具小鋸齒；而絲木棉的葉最為細小，葉緣具鋸齒。木棉的花朵紅艷；吉貝的花朵為黃白色；而絲木棉則為粉色具深色條紋。三者的果實皆為蒴果，但吉貝的果實向上漸狹。

The trees are the members of Bombacaceae. They are commonly found urban tree species in Hong Kong. Without careful observations, it is easy to confuse the three species. We can distinguish them by the trunk, leaves, flowers and fruits.

First, the trunk of Kapok Ceiba is rather spiny, and the basal stem of Silk Floss Tree is distinctively bulging. For the leaves, Tree Cotton contains the largest leaves and the leaf margin is entire. The leaf size of Kapok Ceiba followed that of Tree Cotton and the leaves are minutely serrate. The leaves of Silk Floss Tree are the smallest and is notably serrate. For the flowers, those of Tree Cotton are scarlet, while those of Kapok Ceiba are cream, and those of Silk Floss Tree are pink and with dark stripes. Their fruits are all capsules, but those of Kapok Ceiba are highlighted with tapering apex.

瓜栗 又稱：發財樹

Guinea Peanut, Lucky Tree | *Pachira glabra* Pasq.

相片拍攝地點：獅子會自然教育中心、
新城市廣場對出、九龍公園
Tree Location: Lions Nature Education Centre,
next to New Town Plaza, Kowloon Park

名字由來 MEANINGS OF NAME

　　屬名 *Pachira* 源自蓋亞那的方言，帶有「甜水堅果」（意譯為荸薺）的含意。瓜栗的樹形優美，耐陰能力強，故是一種廣受大眾歡迎的室內植物。其莖部可彎曲，亦可拉伸成各種形狀。在亞洲國家，瓜栗象徵繁榮昌盛，人們相信種植瓜栗可以帶來財富，並能驅邪，故此又被稱為「發財樹」。

本地分佈狀態 DISTRIBUTIONS	外來物種 Exotic species
原產地 ORIGIN	巴西南部和東南部。 South and Southeast Brazil.
生長習性 GROWING HABIT	常綠喬木。高度可達 10 米。 Evergreen tree. Up to 10 m tall.

花果期 月份	1	2	3	4	5	6	7	8	9	10	11	12

花期：本港四月至五月。果期：本港八月至三月。
Flowering period: April to May in Hong Kong. Fruiting period: August to March in Hong Kong.

The generic name *Pachira* is derived from the vernacular language in Guyana, referring to "sweet water nut". Guinea Peanut is a prominent indoor plant in respect of its decorative tree form and excellent tolerance to shade. The stem is malleable and can be stretched into different shapes. In Asian countries, the tree is always symbolized prosperity and the planting of it is believed to bring in wealth and expel evil spirits, hence also named as "Lucky Tree".

應用 APPLICATION

野生瓜栗通常生長在出現季節性澇漬的泛濫平原。其生長速度快，並對乾旱、陰暗及高鹽度環境具高耐受性，故常被種植作林地復興。

The wild Guinea Peanut is prone to grow in floodplains where the lands are always carpeted by seasonal waterlogging. By virtue of its fast growth rate and appreciable adaptability to drought, shade and high salinity, it always serves as a framework species to revitalize forests.

辨認特徵 TRAITS FOR IDENTIFICATION

① 樹幹 TRUNK	② 樹皮 BARK	③ 葉 LEAVES
④ 花 FLOWERS	⑤ 果 FRUITS	

① 瓜栗的樹幹。枝條水平開展。

Trunk of *Pachira glabra* Pasq. Branches horizontally extend.

② 樹皮光滑，綠色。

Bark smooth, green.

③ 掌狀複葉，互生，小葉 5-7 片，近無柄，葉柄兩端具腫脹葉枕。小葉葉片革質，無毛，長橢圓狀倒卵形，頂端漸尖，側脈於葉背隆起。

Palmately compound alternate, leaflets 5-7, nearly sessile, pulvinus swelling at two proximal ends of petiole. Leaflet blade coriaceous, glabrous, oblong-obovate, apex acuminate, lateral veins elevated abaxially.

④ 完全花。單生花或簇生 2-3 朵，腋生。花瓣呈綠色，線狀，向下彎曲。雄蕊眾多，呈白色。

Hermaphroditic. Flowers solitary or 2-3 clustered, axillary. Petals green, linear, recurved. Stamens many, white.

⑤ 蒴果卵形，大型，呈綠色至黃綠色。

Capsules ovoid, large, glabrous, green to yellow green.

植物趣聞 ANECDOTE ON PLANTS

如何分辨瓜栗和馬拉巴栗？ *Pachira glabra* or *Pachira aquatica?*：

瓜栗和馬拉巴栗是兩種不同的物種，然而它們常被世人混淆。根據我們的觀察，本港大部分標示為馬拉巴栗的樹實際上為瓜栗。馬拉巴栗的俗名是美國花生／栗子樹，原生於中美洲和南美洲的熱帶森林。與瓜栗不同的是，野生馬拉巴栗可達 20 米高，樹幹寬闊並具板根。馬拉巴栗的花朵與瓜栗相似，但馬拉巴栗的花絲頂部呈艷紅色，而且果實被褐色毛。

Two trees are all beyond dispute two distinct species. However, people are often confused with the two species. To our observation, most of the trees in Hong Kong labelled *P. aquatica* are actually *P. glabra*. The common name of *P. aquatica* is Malabar Chestnut. It is derived from the tropical forests in Meso and South America. Different from *P. glabra*, the wild Malabar Chestnut can attain a towering height to 20 m, with majestic trunk breadth and buttress roots. Its flowers are similar to that of *P. glabra*, but the filaments' apex is dyed in glamorous red. Instead of glabrous, the fruits are covered with brownish hairs.

備註 Remarks
本樹木學名根據世界植物線上網頁：
Scientific name of this tree is based on Plant of the World Online website :
https://powo.science.kew.org

番木瓜

Papaya | *Carica papaya* L.

相片拍攝地點：柴灣公園、盛泰道
Tree Location: Chai Wan Park, Shing Tai Road

名字由來 MEANINGS OF NAME

屬名 *Carica* 意指卡里亞，起初人們誤認此地為番木瓜屬的原產地。種加詞 *papaya* 源自加勒比印第安語 *ababai*。

The generic name *Carica* refers to Caria that was confused to be the origin of Papaya. The specific epithet *papaya* is derived from the Carib Indian name *ababai*.

應用 APPLICATION

番木瓜是一種在世上被廣泛種植的作物。其肉質果實富含維生素 A、B 和 C 及其他微量營養素如鈣和鐵等，有益身體健康，可直接食用或加工成肉湯。番木瓜的白色乳汁中含有一種蛋白質消化酶 ——「木瓜蛋白酶」，可嫩化肉質。

Papaya is one of the most cultivated crops in the world. The fleshy and wholesome fruits are the abundant sources of vitamins A, B and C, and other micronutrients such as calcium and iron. The fruit can be eaten fresh or processed into soup with meat as the white latex of papaya contains a protein-digesting enzyme "papain" to tenderize meat.

本地分佈狀態 DISTRIBUTIONS	外來物種 Exotic species
生長習性 GROWING HABIT	小型常綠喬木。高度可達 8 米。 Small evergreen tree. Up to 8 m tall.

1	2	3	4	5	6	7	8	9	10	11	12	花果期月份

花期：本港全年可見。果期：本港全年可見。
Flowering period: January to December in Hong Kong. Fruiting period: January to December in Hong Kong.

了解馴化作物的原產地可以為研究進化和人類歷史開闢新思路。然而，野生和考古樣本匱乏，加上長年累月雜交作物以致物種之間關係錯綜複雜，令人們對相關研究一籌莫展。那麼，科學家能否解開這些迷思呢？番木瓜是一個合適的例子用以說明。

番木瓜的起源因缺乏直接的考古證據而一直備受爭議。目前最被廣被接納的假設為「番木瓜原生於南墨西哥至中美洲」。您可能感到疑惑，如果沒有化石記錄，為何科學家能自信地提出這個假設呢？在此之前，我們應先了解一下木瓜的繁殖生物學理論。激素和性染色體決定了植物的性別表達。據我們所知，番木瓜受三種性染色體組合影響，分別是雌性（XX）、雄性（XY）和雌雄同體（XY^h）。Y染色體是雄性性別表達的關鍵，其基因突變可導致由雄性轉性為雌雄同體。至今，雌雄同株的番木瓜在農田中十分普遍。然而，當我們回望過去，古時番木瓜大多數為雌雄異株，雌雄同體的番木瓜猶如一個怪胎。根據近年的分子數據顯示，雄和雌雄同體的番木瓜約在 4000 年前出現分化，這時期與瑪雅文明崛起時期一致。

結合其他間接資訊，包括番木瓜與墨西哥特有種的親緣關係密切，以及墨西哥的番木瓜擁有更高的遺傳多樣性，我們可以粗略推算出以下結論：番木瓜原生於位於中美洲的墨西哥。受到馴化之前，雌雄同體的番木瓜十分稀少。直至約 4000 年前，瑪雅人發現了雌雄同體番木瓜的優點，轉眼間番木瓜便成為了家傳戶曉的水果之一。

Understanding the origins of domesticated crops can open avenues for investigating evolution and human history. However, the paucity of wild and archaeological samples and extremely complicated hybridization events always beleaguer the relative studies. Then, can scientists disentangle these problems? Papaya perhaps is a proper example to elaborate.

The origin of Papaya has been argued for ages due to a dearth of direct archaeological evidence. "Southern Mexico to Central America as the origins of Papaya" is currently the most accepted assumption. You may wonder, if there is no fossil record, how can scientists confidently support the assumption? Before, we should first understand the reproductive biology of Papaya. Plant sex expression is mainly regulated by hormones and sex chromosomes. To our knowledge, Papaya's sex expression is controlled by three types of a pair of sex chromosomes: female (XX), male (XY) and hermaphrodite (XY^h). The Y chromosome is vital for male expression and the genetic mutation of which can shift the male to hermaphrodite. Today, hermaphroditic Papaya is pervasively growing in farmlands. However, when we glimpse into the past, the ancient population of Papaya was actually dominated by distinct males and females, while hermaphrodite was more likely a freak. Supported by the recent molecular data, a divergence between the populations of male and hermaphroditic probably appeared at around 4000 years ago, a time which serendipitously congruent to the rise of the Maya civilization.

Coupled with other indirect information, including the high phylogenetic linkage of Papaya to the plants which are endemic to Mexico, and greater genetic diversity of Papaya showing in Mexico, we can roughly draw a picture as follows: the origin of Papaya was in Mesoamerica, particularly Mexico. Before domestication, the population of hermaphroditic Papaya was rare. Until the Mayan, probably at 4000 years ago, they spotted the advantages of hermaphroditic Papaya and further started the voyage of Papaya to be one of the well-known fruits in the world.

| ① 樹幹 TRUNK | ② 樹皮 BARK | ③ 葉 LEAVES |
| ④ 花 FLOWERS | ⑤ 果 FRUITS | |

① 番木瓜的樹幹。常為單一主莖，具明顯葉痕，或於損傷處分枝，具白色乳汁。

Trunk of *Carica papaya* L. Stem usually solitary, with conspicuous leaf scars, branched when injured, with white latex secretion.

② 樹皮灰白色，光滑。

Bark greyish white, smooth.

③ 葉柄中空，可長達 1 米。單葉互生，常簇生於莖部頂端。葉片近圓形，無毛，掌狀和具 7-11 深裂，每裂片深且具寬鋸齒。

Petioles hollowed, long (to 1 m). Simple leaves alternate, clustered at the apex of stem. Blade suborbicular, glabrous, palmately and deeply 7-11 lobed, each lobe deeply and broadly toothed.

④ 花單性或兩性。雄花無柄，簇生於大型圓錐花序，下垂，花冠筒呈乳白色，雄蕊 10 條，排成 2 輪。雌花具短花梗或無柄，單生或簇生於聚傘花序，花冠筒極短，具 5 裂，子房卵球形，柱頭數裂，近流蘇狀，花冠筒呈乳白色。兩性花簇生於短總狀花序內，雄蕊 5 或 10 條。

Flowers unisexual or bisexual. Male flowers sessile, clustered in a large panicle, pendulous, corolla tube creamy, stamens 10, in 2 whorls. Female flowers with short pedicel or sessile, solitary or aggregated in cymes, corolla tube very short, 5-lobed, ovary ovoid, stigma partite, nearly fimbriate, corolla tube creamy. Bisexual flowers clustered in short racemes, stamens 5 or 10.

⑤ 漿果，倒卵狀球形或橢球形，大而重，成熟時由綠色轉為橙黃色。種子黑色，眾多，具肉質假種皮。

Berries obovoid or ellipsoid, large, heavy, turning from green to orange-yellow when mature. Seeds black, many, with fleshy aril.

枇杷

Loquat │ *Eriobotrya japonica* (Thunb.) Lindl.

相片拍攝地點：樂群街公園、香港動植物公園、獅子會自然教育中心
Tree Location: Lok Kwan Street Park, Hong Kong Zoological and Botanical Gardens,
Lions Nature Education Centre

名字由來 MEANINGS OF NAME

　　種加詞 *japonica* 意指日本，枇杷其實是於唐代由中國傳至日本。由於此品種的葉形酷似中國傳統樂器 —— 琵琶，故又俗稱為「枇杷」。

The specific epithet *japonica* means "from Japan", Loquat was indeed introduced from China to Japan in the "Tang Dynasty". The Chinese common name「枇杷」refers to its pipa-like leaves; pipa is a well-known traditional Chinese musical instrument.

本地分佈狀態 DISTRIBUTIONS	**外來物種** Exotic species
原產地 ORIGIN	枇杷原生於中國，擁有超過 2000 年的栽培歷史。枇杷最初被引入至日本，其後傳播至歐洲和美洲。 Loquat is originated from the South-Central China and has been long cultivated for over 2000 years. The tree was initially introduced to Japan during the Tang Dynasty and further disseminated to Europe and the Americas.
生長習性 GROWING HABIT	**常綠喬木。高度可達 10 米。** Evergreen tree. Up to 10 m tall.

花果期 月份

1	2	3	4	5	6	7	8	9	10	11	12

花期：本港十月至十二月。果期：本港五月至六月。
Flowering period: October to December in Hong Kong. Fruiting period: May to June in Hong Kong.

① 樹幹 TRUNK	② 樹皮 BARK	③ 葉 LEAVES
④ 花 FLOWERS	⑤ 果 FRUITS	

① 枇杷的樹幹。

Trunk of *Eriobotrya japonica* (Thunb.) Lindl.

② 樹皮光滑，呈棕黃色或灰棕色。小枝密被銹色絨毛。

Bark smooth, brownish yellow or greyish brown. Branchlets densely rusty tomentose, stout.

③ 單葉互生，簇生於小枝末端。葉片橢圓狀長橢圓形至闊長橢圓狀披針形。基部楔形，頂端銳形或漸尖，葉緣上部具疏鋸齒。葉背密被銹色絨毛，葉面具光澤及皺摺，托葉鑿形，葉柄較短或缺失。

Simple leaves alternate, crowded at the apex of branchlets. Blade thick leathery, abaxially densely rusty tomentose, adaxially lustrous, rugose, elliptic-oblong to broadly oblong-lanceolate, base cuneate, apex acute or acuminate, margin remotely serrate apically. Petioles short or sessile.

④ 雌雄同株。圓錐花序頂生，密被銹色絨毛。花朵眾多，芳香，花瓣 5 片，呈白色，雄蕊眾多。

Hermaphroditic. Panicles terminal, densely rusty tomentose. Flowers many, fragrant, hypanthium obconical, petals 5, white, stamens many.

⑤ 梨果黃色或橙黃色，球狀或卵狀，具光澤。

Pomes yellow or orange, globose to ovoid, lustrous.

應用 APPLICATION

枇杷具多種經濟和藥用價值，在中國被視為其中一種極其重要的果樹。時至今日，中國依舊是世上最大的枇杷生產國，每年約出口 100 萬噸枇杷。枇杷可直接食用，其肉汁豐富，酸甜可口，深受大眾喜愛。經加工後，可製成枇杷膏，具止咳潤肺、紓緩喉嚨不適之效。枇杷葉味苦澀，用於治療煩躁不安、肺熱燥咳，具有生津止渴之效。

枇杷樹幹直立，樹冠茂密，常被用作觀賞樹種。無憂樹在印度佛教寺廟中被視作神木，並廣泛種植在寺廟內，而枇杷在一些中式佛教寺廟中成為無憂樹的替代品。

Loquat is an excellent fruit tree in China and has been valued for its commercial and medical importance. Today, China is still the leading producer of Loquat fruits in the world, with a stunning one million tons per year. The fruits can be eaten fresh and are beloved for its wholesomeness, sweet and sour tastes, and profuse textures from crispness to juiciness. The fruits can be also processed into loquat paste that can soothe lungs and throat. The leaves are bitter and can be used for treating feverish dysphoria, hydrodipsia cough with lung heat and quenching thirst.

Loquat serves as an ornamental tree because of its dense foliage and handsome tree form. During the flowering period, the tree is blanketed in graceful white blossoms, rendering parks and gardens an alternative aura of vitality. In some Chinese Buddhist temples, Loquat is also planted as a surrogate tree of *Saraca asoca* (Sorrowless Tree), which is a widely planted spiritual tree in Indian Buddhist temples.

鐘花櫻桃 又稱：山櫻、緋寒櫻

Taiwan Cherry │ *Prunus campanulata* Maxim.

相片拍攝地點：香港動植物公園、鰂魚涌公園、元朗公園
Tree Location: Hong Kong Zoological and Botanical Gardens, Quarry Bay Park, Yuen Long Park

名字由來 MEANINGS OF NAME

由於鐘花櫻桃的花朵呈鐘狀，故被冠名 *campanulata*（拉丁文意譯為鐘狀）。

The specific epithet *campanulata* is to describe its bell-shaped flowers.

應用 APPLICATION

與日本不同的是，香港的氣候不適合櫻花生長，因此香港人對香港種植的櫻花印象模糊。鑑於其花朵令人怦然心動，故被零星種植，瑟縮在公園和花園一隅，獨自美麗。鐘花櫻桃可謂香港最常見的櫻花樹，若想體驗賞櫻，在冬末初春的時候前往元朗公園和嘉道理農場暨植物園不失為一個好選擇。

本地分佈狀態 DISTRIBUTIONS	外來物種 Exotic species
原產地 ORIGIN	中國東南部、海南及台灣地區。 Southeast China, Hainan and Taiwan region.
生長習性 GROWING HABIT	小落葉喬木。高度可達 10 米。 Small deciduous tree. Up to 10 m tall.

1	2	3	4	5	6	7	8	9	10	11	12	花果期月份

花期：本港十二月至三月。果期：本港三月至五月。
Flowering period: December to March in Hong Kong. Fruiting period: March to May in Hong Kong.

Unlike Japan, the climate of Hong Kong is not suitable for growing cherry blossom trees. Therefore, Hong Kong people have a vague impression of locally planted cherry blossom trees. They are sporadically planted in parks and gardens because of its delightful flowers. Taiwan Cherry is one of the common Cherry blossom trees that we can see in Hong Kong. If you want to enjoy the cherry blossoms, you may go to Yuen Long Park and Kadoorie Farm and Botanic Garden during the late winter or early spring.

辨認特徵 TRAITS FOR IDENTIFICATION

① 樹幹 TRUNK	② 樹皮 BARK	③ 葉 LEAVES
④ 花 FLOWERS	⑤ 果 FRUITS	

① 鐘花櫻桃的樹幹。

Trunk of *Prunus campanulata* Maxim.

② 樹皮呈暗褐色，光滑，具光澤，具皮孔。

Bark dark brown, smooth, lustrous, lenticellate.

③ 鱗狀冬芽，卵狀，無毛。葉柄具凹槽，葉柄末端具 2 點腺體。托葉具緣毛，早落。單葉互生，葉片無毛，卵形，橢圓狀卵形至卵狀長橢圓形，頂端漸尖，基部近圓形或心形，葉緣銳薄至微鋸齒。

Winter buds scaly, ovoid, glabrous. Petioles grooved, with 2 glands at apex. Stipules ciliate, caducous. Simple leaves alternate. Blade glabrous, ovate, elliptic-ovate to ovate-oblong, apex acuminate, base nearly round or heart-shaped, margin sharp and thin to slightly serrated.

④ 花單生或甚少傘房狀，通常 2-4 朵簇生於葉腋。花朵鐘形，頂生，紅粉色至朱紅色，花萼暗玫瑰色，鐘狀，苞片褐色，花梗纖細。眾多雄蕊，呈黃色。雌雄同體。

Solitary or few corymbose, usually 2-4 clustering in leaf axils. Flowers bell-shaped, pendulous, red, pink, scarlet, calyx deep rose, campanulate, bracts brown, pedicels slender. Stamens many, yellow. Hermaphroditic.

⑤ 核果卵形，成熟時轉為深紫色。

Drupes ovoid, turning dark purple when mature

植物趣聞 ANECDOTE ON PLANTS

日本的櫻花樹 Cherry blossom trees in Japan：

　　為了增加花朵顏色及形狀豐富度，李屬被人們廣泛栽培／雜交。為了讓鐘花櫻桃的花期連綿不斷，人們通常會通過嫁接繁殖的方法，以確保每顆樹都是母株的複製品。對日本而言，櫻花樹擁有極其重要的文化意義，故被視為國花。櫻花在日語中發音為さくら（sakura）。日本人對櫻花充滿熱情，可謂為之瘋狂。早在平安時代，他們的祖先早已體會到櫻花轉瞬即逝的淒美，並為櫻花賦予多番美譽。自始至今，日本人每年春天都會舉行賞花大會，俗稱賞櫻，日語為花見（はなみ／hamami）。

　　日本人對櫻花十分渴慕，如鹿切慕溪水，故此當地人對培植櫻花興趣盎然。染井吉野櫻是一種由一葉櫻和大島櫻雜交而成的品種，是目前日本最主要的櫻花樹。根據分子數據分析顯示，所有日本的染井吉野櫻皆由同一株母株複製而成，換句話說，它們的基因一致。如果科學家能對其他櫻花樹進行研究，探明當中栽培史的奧祕，便能在植物學上再創一舉。

Prunus spp. has been widely cultivated/hybridized for proliferating the diversity of floral shapes and colours. To maintain ubiquitous blossoming, the trees are always propagated by grafting to ensure every tree is a "clone" of the mother tree. Cherry blossom trees are culturally important and considered as the national flower of Japan. Its blossoms in Japanese is pronounced as *Sakura*. Japanese people are passionate about *Sakura*. Early in the Heian period (794 to 1185), their ancestors already recognized the transient beauty of the blossoms and conferred them multiple praises. Today, Japanese people have pursued the flower-viewing activity in every spring, well-known as *Hanami*.

With a great appetite for the blossoms, Japanese people have excitedly engaged in cultivating Cherry blossom trees. *Prunus × yedoensis*, a hybrid of *Prunus lannesiana* and *Prunus speciosa*, is currently the most dominant Cherry blossom tree that has spread throughout Japan. Supported by the recent molecular data, all *Prunus × yedoensis* in Japan could be cloned from the same mother tree. In other words, they are genetically the same. It would be exciting if scientists can endeavour on studying if other Cherry blossom trees are sharing the comparable phenomenon and soon ascertain the comprehensive history of cultivation of these marvellous plants.

海紅豆 又稱：孔雀豆、相思格

Red Sandalwood | *Adenanthera microsperma* Teijsm. & Binn.

相片拍攝地點：九龍公園、荔枝角公園
Tree Location: Kowloon Park, Lai Chi Kok Park

名字由來 MEANINGS OF NAME

屬名 *Adenanthera* 源自希臘語單詞 *aden*（黏性腺體）和 *anthera*（花藥），共同指花藥上的微小腺體。海紅豆的種子與紅豆十分相似，故有「海紅豆」此中文名稱。

The generic name *Adenanthera* is derived from the Greek words *aden* (sticky gland) and *anthera* (anthers), together alluding to tiny glands on the anthers. The Chinese name「海紅豆」describes its seeds are red-beans-like.

應用 APPLICATION

海紅豆的種子用於製作手鐲以表愛意。全株有毒，很少口服。它的種子通常被磨成粉末外用，治療臉上的黑斑、痤瘡和紅斑痤瘡。

The seeds are always processed into bracelets as emblematic of love. The entire plant is toxic and seldom taken orally. Instead, the seeds can be grounded into powder for external treatments like expelling dark spots on face, acne and brandy nose.

本地分佈狀態 DISTRIBUTIONS	原生物種 Native species
原產地 ORIGIN	雲南、貴州、廣西、廣東、福建、海南和台灣。緬甸、柬埔寨、老撾、越南、馬來西亞、印尼也有分佈。 Yunnan, Guizhou, Guangxi, Guangdong, Fujian, Hainan and Taiwan. Also Myanmar, Cambodia, Laos, Vietnam, Malaysia and Indonesia.
生長習性 GROWING HABIT	落葉喬木。高度可達 20 米。 Deciduous tree. Up to 20 m tall.

花果期 月份	1	2	3	4	5	6	7	8	9	10	11	12

花期：本港四月至七月。果期：本港七月至十月。
Flowering period: April to July in Hong Kong. Fruiting period: July to October in Hong Kong.

① 樹幹 TRUNK	② 樹皮 BARK	③ 葉 LEAVES
④ 花 FLOWERS	⑤ 果 FRUITS	

① 海紅豆的樹幹。

Trunk of *Adenanthera microsperma* Teijsm. & Binn.

② 樹皮灰色，光滑，稍微薄鱗片狀剝落。小枝被微柔毛。

Bark grey, smooth, slightly peeling off in thin scales. Branchlets puberulent.

③ 二回羽狀複葉對生，羽片 3-5 對，對生或近對生，小葉 4-7 對，互生。葉片長橢圓形或卵形，兩端圓鈍狀，兩面皆被微柔毛，背面淡綠色。葉柄較短，複葉總柄呈紅棕色，表面具縱向淺溝。

Bipinnately compound alternate, pinnae 3-5 pairs, subopposite, leaflets 4-7 pairs, alternate. Petioles and rachis pubescence. Blade oblong or ovate, broadly obtuse at both ends, puberulent on both sides, light green on back. Petiole shorter, compound rachis reddish brown, surface with longitudinal shallow groove.

④ 總狀花序腋生或圓錐花序頂生，呈金黃色。

Racemes axillary or panicles terminal, golden-yellow or white.

⑤ 莢果狹長橢圓形，開裂後果瓣旋卷，露出紅色具光澤的扁圓形種子。

Pods narrowly oblong, valves contorted when dehiscent, exposing red and glossy oblate seeds.

生命力 VITALITY

海紅豆是一種快速生長品種，偏好生長於酸性和潮濕的土壤。

The tree is fast-growing and prefers acidic and moist soils.

植物趣聞 ANECDOTE ON PLANTS

海紅豆和紅豆 The imaginations to Red Sandalwood：

我們的祖先十分鍾愛紅豆，他們喜愛用紅豆來表達他們的愛和情懷。相傳，有一對夫婦居住在越國的鄉村。有一日，婦人的丈夫被迫駐守邊疆，此後便沒有再回家。婦人堅持要在村前的樹下等待丈夫歸來，哭得眼睛都流血了。最終，婦人哭死在樹下，樹上亦結滿了紅豆。之後，人們相信婦人流下的眼淚化為一顆顆紅豆，掛滿樹上，遂把此樹命名為「相思子」。紅豆亦被廣泛引用於詩詞歌賦中，例如王維的詠物詩《紅豆》也引用了紅豆，以比喻表達他的情懷。然而，人們曾爭論文獻中指出的「紅豆」到底是甚麼品種，答案五花八門，相思子、海紅豆和紅豆樹都可能是當時文獻所指的品種，但是答案仍然備受爭議，未有定論。

Our ancestors were an aficionado of red beans and used them to illustrate their love and longing. Rumour has it that a woman lived with her husband in a village in Yue State. One day, her husband was forced to garrison the frontiers and then never returned home. Because of yearning, she insisted on waiting for her husband under a tree in front of the village and cried until her eyes got bleeding. After the woman died with persistent crying, the tree just astonishingly yielded numerous red beans. Since people believed that the seeds were derived from the woman's tears, and thus named the tree as「相思子」to commemorate this miserable love story. Red beans were widely cited in the poem, such as in Wang Wei's poem *Red Bean*. However, the suggested species that "red beans" in the literature, to whether Red Sandalwood, *Abrus precatorius* (Rosary Pea) or *Ormosia hosiei* (Hainan Ormosia) it referred, is still controversial.

合歡

Silk-tree, Mimosa │ *Albizia julibrissin* Durazz.

相片拍攝地點：迪欣湖活動中心、西九
文化區
Tree Location: Inspiration Lake Recreation Centre,
West Kowloon Cultural District

名字由來 MEANINGS OF NAME

　　為紀念意大利博物學家 —— 艾菲利波·德利·阿爾比齊於 1749 年引入合歡
至意大利的托斯卡納，屬名特此取其名為 *Albizia*。種加詞 *julibrissin* 源自波斯語
Ghulibrechim，暗指其呈粉紅色和纖毛狀的雄蕊（波斯語意譯為絲綢玫瑰）。故合歡
亦有「絨花樹」的別名。

本地分佈狀態 DISTRIBUTIONS	**外來物種** Exotic species
原產地 ORIGIN	**亞洲及非洲熱帶。** Asia and tropical Africa.
生長習性 GROWING HABIT	**落葉喬木。高度可達 16 米。** Deciduous tree. Up to 16 m tall.

1	2	3	4	5	6	7	8	9	10	11	12

花果期
月份

花期：本港五月至七月。果期：本港八月至十月。
Flowering period: May to July in Hong Kong. Fruiting period: August to October in Hong Kong.

135

The generic name *Albizia* commemorates an Italian naturalist, Fillippo delgi Albizzia, who first introduced the tree into Tuscany in Italy in 1749. The specific epithet *julibrissin* comes from the Persian word *Ghulibrechim* (silk-rose), alluding to its pink and ciliate stamens. As a result, the tree is also named as "Silk-tree".

辨認特徵 TRAITS FOR IDENTIFICATION

① 樹幹 TRUNK　② 樹皮 BARK　③ 葉 LEAVES
④ 花 FLOWERS　⑤ 果 FRUITS

① 合歡的樹幹。

Trunk of *Albizia julibrissin* Durazz.

② 樹皮淡灰色，光滑，具皮孔。小枝具棱角。

Bark light grey, smooth, lenticellate. Branchlets angular.

③ 二回羽狀複葉，互生，羽片 4-12 對，對生。小葉 10-30 對，對生，無柄，斜線形至長圓形，基部截形，頂端細尖，葉緣具毛，中脈靠近上葉緣。腺體位於最上端羽片的葉軸和葉柄基部。

Bipinnately compound alternate, pinnae 4–12 pairs, opposite, leaflets 10–30 pairs, opposite, sessile, obliquely linear to oblong, base truncate, apex apiculate, margin ciliate, midvein close to the upper edge. Glands at leaf rachis of the uppermost pinnae and the proximal end of petioles.

④ 圓錐花序，頂生，15-20 朵簇生於頭狀花序，於枝頂排成圓錐花序。花朵芳香，呈粉紅色，雄蕊眾多。

Panicles terminal, 15-20 clustered in head, peduncles pubescent. Flowers fragrant, pink, with many stamens.

⑤ 莢果帶狀，扁平，成熟時轉為褐色。
Pods strap-shaped, flat, brown when ripe.

應用 APPLICATION

合歡木材堅硬，故常被當地人用於製作家具和建材。其樹皮提取物具抗炎作用，可用於緩解肺部疼痛，治療傷口和瘀傷，並能減輕失眠、精神錯亂和乏力等狀況，促進心理健康。種子油的提取物含有豐富脂肪酸，故常加工成洗髮乳和香皂。

合歡觀賞價值高，故一直被視為觀賞樹種栽培。根據記錄顯示，合歡早在 18 世紀已被引入至英國、意大利和北美洲。合歡亦在許多中國文學作品中被提及。以「詩聖」杜甫（712-770）所作的《佳人》為例，詩句「合昏尚知時，鴛鴦不獨宿」中的「合昏」意指合歡。

在香港，此樹種作為少見的觀賞樹種，只有少數種植在西九文化區和香港迪士尼樂園。當合歡盛開時，一朵朵粉撲般的嬌花掛滿枝頂，吸引遊客佇足觀賞。其花芳香四溢，對傳粉者極具吸引力，令傳粉者蜂擁而至，吸食其花蜜，為合歡繁殖出一分力。

The wood is stiff and highly demanded for making furniture and other constructions in its native range. The bark extract is anti-inflammatory that is effective to relieve lung pain, wounds and bruises, and promote mental health by attenuating insomnia, confusion and asthenia. Since the seed oil contains profuse fatty acids, it is always processed into hair shampoos and soaps.

Silk-tree has been long cultivated by virtue of its spectacular ornamental value, with the early introduction events since the 18th century in England, Italy and North America. The tree was also mentioned frequently in Chinese literature. For instance,「合昏」mentioned in the poetry《佳人》by Du Fu (712-770),「合昏尚知時，鴛鴦不獨宿」, could probably refer to「合歡」.

This ornamental tree is not common in Hong Kong, with individuals scattering in West Kowloon Cultural District and Hong Kong Disneyland. The tree is not valuable for its green foliage; instead, when it flowers, the glamourous pink blossoms festoon the crown and are highly ornamental, attracting visitors to stand and appreciate. The fragrant smell of the flowers is very attractive to pollinators. The pollinators will come to the flower and forage the nectar, assisting the reproduction of the tree.

生命力 VITALITY

合歡樹生長迅速，耐寒且耐旱，故一直被視為沙地和不毛之地的先鋒物種。

Silk-tree is fast-growing and hardy to cold and drought, hence usually regarded as a pioneer species in sandy and barren sites.

南洋楹

Batai Wood | *Falcataria moluccana* (Miq.) Barneby & J. W. Grimes

相片拍攝地點：港鐵上水站對出
Tree Location: Next to Sheung Shui MTR Station

名字由來 MEANINGS OF NAME

屬名 *Falcataria* 為意大利文中「鐮刀」的意思，暗指其鐮狀葉形。種加詞 *moluccana*，意指其原生於摩鹿加群島。

The generic name *Falcataria* means falcate, referring to its sickle-like leaflet shape. The specific epithet *moluccana* depicts its origin from the Moluccas.

本地分佈狀態 DISTRIBUTIONS	外來物種 Exotic species
原產地 ORIGIN	原生於印尼（爪哇島、蘇門答臘、摩鹿加群島和伊里安查亞）、馬六甲、巴布亞新幾內亞（布干維爾島和俾斯麥群島）和所羅門群島。 Native to Indonesia (Java, Sumatra, Moluccas and Irian Jaya), Malacca, Papua New Guinea (Bougainville island and Bismarck Archipelago) and the Solomon Islands.
生長習性 GROWING HABIT	常綠喬木。高度可達 45 米。傘狀樹形。 Evergreen tree. Up to 45 m tall. Umbrella-like tree.

花果期 月份	1	2	3	4	5	6	7	8	9	10	11	12

花期：本港四月至七月。果期：本港九月至十二月。
Flowering period: April to July in Hong Kong. Fruiting period: September to December in Hong Kong.

南洋楹生長迅速，木質較軟，較易砍伐，故被認為是理想木材樹種。其木材常被用作木柴或加工成卡板、包裝盒、筷子和樂器。

成熟的南洋楹的樹冠格外緊密，能提供優越的遮蔭效果，並緩解市區環境問題，如熱島效應。由於其樹形壯大，枝條易脆。為減少因狂風造成南洋楹倒塌的情況，應將其種植在開揚的空地上，以避免對途人構成危險。

Since the tree is fast-growing and the wood is relatively easy to chop, it is regarded as an ideal timber tree. The wood serves versatile functions, mostly treated as fuelwood or processed into pallets, packing boxes, chopsticks and musical instruments.

The exceptionally extensive and compact crown enables the tree to provide notable shading effect. It is also competent to attenuate urban environmental problems such as the heat island effect. In respect of its spectacular tree form but relatively fragile branches, it should be planted on wide open space with enough protection from wind to avert compromise to the safety of human property.

辨認特徵 TRAITS FOR IDENTIFICATION

① 樹幹 TRUNK	② 樹皮 BARK	③ 葉 LEAVES
④ 花 FLOWERS	⑤ 果 FRUITS	

① 南洋楹的樹幹。

Trunk of *Falcataria moluccana* (Miq.) Barneby & J. W. Grimes.

② 樹皮淡灰色，光滑。小枝稍被短柔毛，具皮孔。

Bark light grey, smooth. Branchlets minutely pubescent, lenticellate.

③ 二回羽狀複葉，互生，羽片 6-20 對，小葉 6-26 對，無柄，對生，近小葉和羽片基部具盤狀腺體。小葉紙質，斜長圓形、鐮刀形、基部鈍形、圓形或近楔形，頂端銳形，中脈偏近葉片葉片上部，基出脈 2-3 條。

Bipinnately compound alternate, pinnae 6-20 pairs, pinnules 6-26 pairs, opposite, sessile, disk-shaped glands near the base of pinnules and pinnae. Blade papery, obliquely oblong, falcate, base obtuse, round or nearly cuneate, apex acute, midvein towards the upper blade, basal veins 2-3.

④ 雌雄同株。穗狀花序，腋生，花單生或組成圓錐狀花序。花朵呈黃綠色至奶油色，花萼寬鐘狀，雄蕊眾多。

Hermaphroditic. Spikes axillary, solitary or forming a panicle. Flowers greenish yellow to cream, calyx broadly campanulate, stamens many.

⑤ 莢果直而扁平，成熟時轉為褐色並裂開。

Legumes straight, flat, turning brown and dehiscent when ripe.

植物趣聞 ANECDOTE ON PLANTS

南洋楹的入侵性 Invasiveness of Batai Wood：

南洋楹在陽光普照、土壤乾燥的情況下，根系仍能急速發展，加上其生長速度極快，故種植作復育不毛之地。從某種意義上，南洋楹可說是一把雙刃劍，因上述特徵令其有高入侵性，這樹因嚴重入侵太平洋島嶼而臭名昭著。其具有固氮細菌的特徵能協助將氮固定成植物可吸收的狀態，這能令南洋楹能與島上其他原生物種一決高下。其發達的根系令它在爭取天然資源，比起其他原生種更勝一籌；龐大的樹冠限制了原生物種的繁衍。

The tree prospects in sunny weather and dry soil due to its extensive roots. Since the tree can grow rampantly fast, it is always planted for revitalizing barren lands. In a sense, the tree also serves as a double-edged sword with the above characteristics as leading the tree to have a high invasiveness. Batai Wood is notorious in Pacific Islands for its dramatic intrusion into the local environments. The tree equips nitrogen-fixing bacteria and is able to fix the unavailable nitrogen into plant-available status, which renders the tree outstanding compatibility beyond other native species in the islands. The extensive root system also flavours it from the competition of resource. The colossal tree crown blocks the sunlight and suppresses the growth of native seedlings.

儀花

Lysidice | *Lysidice rhodostegia* Hance

相片拍攝地點：九龍公園、香港動植物公園、雷公田
Tree Location: Kowloon Park, Hong Kong Zoological and Botanical Gardens, Lui Kung Tin

應用 APPLICATION

儀花迷人的花朵在夏日綻放，為公園和花園營造怡人景色。其茂密的樹幹能提供優良的遮蔭效果，故亦被視為遮蔭樹廣泛種植。

人們把它的根挖出來，並曬乾以供藥用。其味苦辛，具活血、消腫、止血之效。由於儀花的根部、莖部和葉片皆帶有輕微毒素，入藥前應先諮詢醫生的專業意見。

Lysidice is a graceful greening component to parks and gardens in respect of its glamorous blossoms in summer. In addition, the dense foliage renders the tree excellent shading effect and therefore is widely planted as shading trees.

The roots are excavated and dried for multiple medicinal purposes. It tastes bitter and pungent, and is well-known for the promotion of blood circulation, detumescence and hemostasis. Since the roots, stems and leaves of Lysidice are mildly toxic, the prescription for medicines should be dispensed by doctors.

本地分佈狀態 DISTRIBUTIONS	外來物種 Exotic species
原產地 ORIGIN	中國中南部、東南部及越南。 South-Central and Southeast China, and Vietnam.
生長習性 GROWING HABIT	落葉灌木或喬木。高度通常 5 米，甚少達 10 米。 Deciduous shrub or tree. 5 m tall, rarely taller than 10 m.

1	2	3	4	5	6	7	8	9	10	11	12	花果期 月份

花期：本港六月至八月。果期：本港九月至十一月。
Flowering period: June to August in Hong Kong. Fruiting period: September to November in Hong Kong.

① 樹幹 TRUNK	② 樹皮 BARK	③ 葉 LEAVES
④ 花 FLOWERS	⑤ 果 FRUITS	

① 儀花的樹幹。

Trunk of *Lysidice rhodostegia* Hance.

② 樹皮光滑，呈灰棕色至黃棕色。

Bark smooth, greyish brown to yellowish brown.

③ 羽狀複葉，互生，小葉 3-5 對，對生。葉柄基部具葉枕，葉柄較短。小葉葉片長橢圓形或卵狀披針形，基部鈍形，頂端尾形至漸尖，葉緣全緣。側脈細長且密集。

Pinnately compound alternate, leaflets 3-5 pairs, opposite. Pulvinus at the proximal end of petiole, petiolule short. Blade long elliptic or ovate-lanceolate, base obtuse, apex caudate to acuminate, entire, lateral veins slender, dense.

④ 完全花。圓錐花序頂生或腋生。苞片及小苞片呈白色至粉紅色，卵狀長橢圓形或橢圓形。萼管長於萼裂片，4 裂，呈淺紫紅色。花瓣 5 片，其中 3 片明顯，呈暗紫紅色，爪狀，另外 2 片較小，呈鱗片狀。

Hermaphroditic. Panicles terminal or axillary. Bracts and bracteoles white to pink ovate-oblong or elliptic. Calyx lobes 4, shorter than calyx tube, light purplish red. Petals 5, only 3 of them notable, dark purplish red, clawed, another 2 small, scale-like.

⑤ 莢果，倒卵狀長橢圓形，密集，基部稍傾斜，背縫線長短不等，6-10 粒種子，成熟時變乾，由綠色轉為褐色。種子棕紅色，長橢圓形，密集。

Pods, obovate-long-elliptic, dense, base slightly tilted, dorsal sutures varying in length, 6-10-seeded, drying at maturity, turning from green to brown. Seeds brownish red, long elliptic, dense.

植物趣聞 ANECDOTE ON PLANTS

儀花與短萼儀花 *Lysidice rhodostegia* and *L. brevicalyx*：

儀花與短萼儀花的外形大同小異。要區分兩者，儀花的萼管較小苞片長，而短萼儀花的萼管則較儀花短，與小苞片長度相若。

Lysidice is morphologically identical to *L. brevicalyx*. To distinguish the trees, Lysidice's flowers carry a longer calyx tube which is much longer than bracteoles, while the calyx tube of *L. brevicalyx* is shorter and as long as bracteoles.

古樹名木 Old and Valuable Tree（OVT）：

儀花甚少長成參天巨木。在香港動植物公園園內，有一棵儀花被列入古樹名木（編號：LCSD CW/58），其高約 18 米，胸徑約 950 毫米，樹冠約闊 15 米。

Lysidice rarely grows into a majestic size. In Hong Kong Zoological and Botanical Gardens, a Lysidice is registered as a OVT (LCSD CW/58), with its towering height of 18 m, diameter at breast height (DBH) of 950 mm, and opening crown spread of 15 m.

儀花古樹名木

降香黃檀

Fragrant Rosewood | *Dalbergia odorifera* T. C. Chen

相片拍攝地點：沙田公園、大埔海濱公園、香港大學
Tree Location: Sha Tin Park, Tai Po Waterfront Park, The University of Hong Kong

名字由來 MEANINGS OF NAME

　　為紀念「瑞典兄弟」—尼古拉斯·達爾伯格和卡爾·古斯塔夫·達爾伯，所以取屬名 *Dalbergia*。前者是瑞典植物學家，後者因參與探索西印度群島而聞名。切割降香黃檀的木材時，會散發出陣陣木香，芳香四溢，故被冠名為 *odorifera*（中文意譯為香氣），亦有 Fragrant Rosewood 的俗名（中文意譯為降香黃花梨）。

本地分佈狀態 DISTRIBUTIONS	**外來物種** Exotic species
原產地 ORIGIN	**海南。** Hainan.
生長習性 GROWING HABIT	**半落葉喬木。高度可達 15 米。** Semi-deciduous tree. Up to 15 m tall.

花果期 月份	1	2	3	4	5	6	7	8	9	10	11	12

花期：本港四月至六月。果期：本港七月至十二月。
Flowering period: April to June in Hong Kong. Fruiting period: July to December in Hong Kong.

The generic name *Dalbergia* commemorates Swedish brothers Nicholas Dalberg (1736-1820) and Carl Gustav Dalberg (1721-1781). The former was botanist and the later was well-known for his engagement in the exploration of the West Indies. The specific epithet *odorifera* refers to its fragrant wood when chopped. By the same reason, the tree is also named as "Fragrant Rosewood".

應用 APPLICATION

降香黃檀擁有無可比擬的經濟和藥用價值，所以人們都夢寐求之。其心材在中國被稱為「花梨木」或「黃花梨」。花梨木的心材色澤黃潤，花紋絢麗，氣味芳香，經久耐用，穩定性高，是製作家具和工藝品的優良木材。花梨木生長緩慢，長達50年以上才踏入成熟期，因此成本高昂，被視為一種名貴的木材。

降香黃檀的心材是一種極其重要的中藥材，名為「降香」。降香含有豐富的類黃酮和酚類等副產物，有不少研究證實能夠活血化瘀、紓緩癌症、缺血和風濕痛症等。

Fragrant Rosewood has been yearned for its unsurpassed economical and medicinal values. Its heartwood is named as *Hualimu* or *Huanghuali* in China. *Hualimu* is an excellent timber for making furniture and crafts by virtue of its glamorous colour and pattern, aromatic smell, respectable durability and stability. By reason of its sluggish growing rate which requires 50 years for attaining a mature size, Hualimu is regarded as an extravagant wood and the cost is generally unaffordable.

The heartwood is also the primary source of a traditional Chinese medicine, known as *Jiangxiang*. *Jiangxiang* contains profuse secondary products such as flavonoids and phenolic. The medicinal effects have been well-studied for relieving blood disorders, cancer, ischemia and rheumatic pain.

辨認特徵 TRAITS FOR IDENTIFICATION

① 樹幹 TRUNK	② 樹皮 BARK	③ 葉 LEAVES
④ 花 FLOWERS	⑤ 果 FRUITS	

① 降香黃檀的樹幹。

Trunk of *Dalbergia odorifera* T. C. Chen.

② 樹皮褐色至淡褐色，粗糙，縱裂。小枝具皮孔。

Bark brown to pale brown, rough, longitudinally splitting. Branchlets lenticellate.

③ 奇數羽狀複葉，互生，小葉 7-13 片，互生。葉柄基部具腫脹葉枕，葉柄較短。小葉片近革質，卵形或橢圓形，基部圓形或寬楔形，頂端鈍形至漸尖。

Imparipinnately compound, alternate, leaflets 7-13, alternate. Pulvinus swelling at the proximal end of petioles, petiolules short. Blade thinly leathery, ovate or elliptic, base rounded or broadly cuneate, apex obtuse to acuminate.

④ 雌雄同體。圓錐花序腋生。花朵較小，呈白色或淡黃色，花萼鐘狀，花瓣爪形。

Hermaphroditic. Panicles axillary. Flowers small, white or pale yellowish, calyx campanulate, petals clawed.

⑤ 豆莢長橢圓形，先端鈍形或銳形，基部突變窄至細長，變乾時具明顯網狀紋，1 或 2 粒種子。種子腎形，密集。

Legumes oblong, apex obtuse or acute, base abruptly narrowed to slender, notably reticulated when dried, 1 or 2 seeded. Seeds reniform, compressed.

植物趣聞 ANECDOTE ON PLANTS

易危物種 Vulnerable species：

降香黃檀價值連城，惹來人們濫伐，導致原生地的野生降香黃檀數量大幅減少。降香黃檀的遺傳變異遠低於預期，難以適應環境變化。從保育生物學而言，如果沒有適當的保育策略，降香黃檀很容易會走上滅絕之路。故此，降香黃檀被國際自然保護聯盟（IUCN）列為「易危」，同時在香港目前受到第 586 章《保護瀕危動植物物種條例》保護約束。未經許可，嚴禁對降香黃檀作出任何破壞行為或進行交易。

In respect of its values, the wild Fragrant Rosewood has been rampantly chopped and the wild population has been significantly declined. According to the results of some genetic studies, the genetic variation of the wild Fragrant Rosewood is far lower than expected and could be extremely fragile to any environmental change. From the perspective of conservation biology, the tree can easily drift to extinction without any proper conservative strategy. As a result, Fragrant Rosewood is rated as Vulnerable by the International Union for Conservation of Nature (IUCN) while the timber trading is regulated by the Protection of Endangered Species of Animals and Plants Ordinance (Cap. 586) in Hong Kong. Without permit, any trading of the tree is forbidden.

海南紅豆

Hainan Ormosia | *Ormosia pinnata* (Lour.) Merr.

相片拍攝地點：九龍公園、大埔海濱公園、夏慤花園
Tree Location: Kowloon Park, Tai Po Waterfront Park, Harcourt Garden

名字由來 MEANINGS OF NAME

屬名 *Ormosia* 源自希臘語 *hormos*，帶有鏈條或項鍊的含意，暗指其種子形似串成項鍊的珠子。海南紅豆的葉片為奇數羽狀複葉，故被冠名為 *pinnata*（中文意譯為羽毛狀）。

The generic name *Ormosia* is derived from the Greek word *hormos* (a chain or necklace), alluding to its necklace-like seeds. The specific epithet *pinnata* refers to its imparipinnately compound leaves.

本地分佈狀態 DISTRIBUTIONS	外來物種 Exotic species
原產地 ORIGIN	廣東、海南、廣西。 Guangdong, Hainan and Guangxi.
生長習性 GROWING HABIT	常綠喬木。高度可達 30 米。 Evergreen tree. Up to 30 m tall.

1	2	3	4	5	6	7	8	9	10	11	12	花果期月份

花期：本港七月至八月。果期：本港十月。
Flowering period: July to August in Hong Kong. Fruiting period: October in Hong Kong.

在陽光充沛的自然生境中，海南紅豆便能呈現優美的樹形及長出茂密的樹冠。因此，此樹種既適合單獨種植，亦可與公園或街道上的其他植物合併種植，以提供絕佳的遮蔭效果。海南紅豆奇特的花和種子亦令此樹種備受重視。百花盛開之時，無數傳粉者悄悄造訪，為城市賦予生機。

其心材顏色呈迷人的淺紅棕色，而且不易腐爛，故常被用於家具製作和建材。鮮艷亮麗的紅色種子可用作裝飾用途。

In its natural range with sufficient sunlight support, Hainan Ormosia can exhibit a handsome tree form with dense foliage. Therefore, good planting should be conducted with full sunlight exposure. The tree can be either planted in solitary or mixed planting into other vegetative configurations in parks and streets for providing an unsurpassed shading effect. The tree is also valued for its remarkable flowers and seeds. When it blossoms, numerous pollinators tap the flowers and fill the urban jungle with vigour and vitality.

In respect of its appealing heartwood colour in light reddish-brown and relatively resistant to rotting, its wood is always used for making furniture and construction. The seeds dyed in a fascinating red colour can be used for decorations.

辨認特徵 TRAITS FOR IDENTIFICATION

① 樹幹 TRUNK	② 樹皮 BARK	③ 葉 LEAVES
④ 花 FLOWERS	⑤ 果 FRUITS	

① 海南紅豆的樹幹。

Trunk of *Ormosia pinnata* (Lour.) Merr.

② 樹皮灰色至棕灰色，光滑。小枝被褐色毛狀體，成熟時轉為無毛。

Bark grey to brownish grey, smooth. Branchlets covered with brown trichomes, turning glabrous when aging.

③ 奇數羽狀複葉，互生，小葉 7-11 片，對
生，頂生小葉具長葉軸。葉柄基部具葉枕，
葉柄具凹槽，被短柔毛。小葉片披針形至
倒披針形，薄革質，無毛，頂端鈍形或銳
形，側脈 5-7 對。

Imparipinnately compound alternate, leaflets
7-11, opposite, the terminal leaflet with elongated
rachis. Pulvinus at the proximal end of petioles,
petiolules grooved, pubescent. Blade lanceolate
to oblanceolate, thin-leathery, glabrous, apex
obtuse or acute, lateral veins 5-7.

④ 圓錐花序頂生，被褐色短柔毛。雌雄同
體，花萼鐘狀，邊緣具鋸齒，寬三角形，
花冠呈乳黃色。

Hermaphroditic. Terminal panicles, brown
pubescent. Calyx campanulate, margin toothed,
broad-triangular, corolla cream.

⑤ 莢果稍彎曲，花柱宿存，裂片木質。種
子橢圓形，珠柄短小，被紅色假種皮包圍。

Legumes slightly curved, with persistent style,
valves woody. Seeds ellipsoid, enclosed by red
aril, funiculus short.

昆士蘭銀樺 又稱：紅花銀樺
Scarlet Grevillea, Bank's Grevillea | *Grevillea banksii* R. Br.

相片拍攝地點：鰂魚涌公園
Tree Location: Quarry Bay Park

名字由來 MEANINGS OF NAME

　　屬名 *Grevillea* 是為了紀念查爾斯・法蘭西斯・格雷維爾（1749-1809）。他為了成立英國皇家園藝學會而鞠躬盡瘁。種加詞 *banksii* 是為了表彰曾任皇家學會會長的約瑟夫・班克斯（1743-1820）。中文名稱「昆士蘭」和「紅花」分別源自其原產地和紅色的花色。

　　The generic name *Grevillea* is named after Charles Francis Greville (1749-1809), who participated hardly in the establishment of the Horticultural Society (currently known as the Royal Horticultural Society) in the Britain. The specific epithet *banksii* is in honour of Joseph Banks (1743-1820), who was the President of the Royal Society. The Chinese common names「昆士蘭」and「紅花」refer to its origin and red blossoms, respectively.

本地分佈狀態 DISTRIBUTIONS	外來物種 Exotic species
原產地 ORIGIN	澳洲昆士蘭州。 Queensland , Australia.
生長習性 GROWING HABIT	常綠灌木或喬木。高度可達 10 米。 Evergreen shrub or tree. Up to 10 m tall.

花果期月份

1	2	3	4	5	6	7	8	9	10	11	12

花期：本港一月至六月。果期：本港十二月至二月。
Flowering period: January to June in Hong Kong. Fruiting period: December to February in Hong Kong.

　　昆士蘭銀樺擁有讓人神搖意奪的花朵，故作為觀賞樹種被廣泛種植在公園和花園中。請謹記，昆士蘭銀樺不可食用。它的花朵和果實中含有大量具毒性的氰化物。

In respect of its glamorous blossoms, the tree is widely cultivated in parks and gardens for ornamental purposes. Notably, Scarlet Grevillea is inedible. Its flowers and fruits contain abundant cyanogenic substances which are toxic.

辨認特徵 TRAITS FOR IDENTIFICATION

① 樹幹 TRUNK	② 樹皮 BARK	③ 葉 LEAVES
④ 花 FLOWERS	⑤ 果 FRUITS	

① 昆士蘭銀樺的樹幹。

Trunk of *Grevillea banksii* R. Br.

② 樹皮褐色，縱裂。幼枝及花序被銀灰色或銹色柔毛。

Bark dark brown, longitudinally cracked. Young branchlets covered with silver hairs.

③ 單葉互生，羽狀深裂，4-12 片裂片，線狀，葉面被柔毛或無毛，葉背被銀白色絨毛，葉背葉緣稍卷曲。

Simple leaves alternate. Blade pinnatiparted, segments 4-12, linear, adaxially hairy or glabrous, abaxially covered with silver hairs, margin slightly recurved.

④ 頂生總狀花序，花朵呈紅色，花柱外露。

Terminal racemes. Flowers scarlet, style protruded.

⑤ 蓇葖果，初時綠色被絨毛，成熟時轉為褐色。

Follicles tomentose, turning from green to brown when mature.

昆士蘭銀樺的花朵和果實對鳥類和蝴蝶具高吸引力。然而，它會釋出一些植化相剋的化感物質，這些物質導致土壤微生物的活動中斷，抑制當地植物的繁衍，故在馬達加斯加等國家中被歸類為入侵性品種。

The flowers and fruit of Scarlet Grevillea are highly attractive to birds and butterflies. However, Scarlet Grevillea will release some allelochemicals, which interrupt the activities of soil microorganisms and inhibit the reproduction of local plants, so they are classified as invasive species in Madagascar and other countries.

生命力 VITALITY

昆士蘭銀樺對陰暗環境具高耐受性，但偏好全日照的生長環境和排水良好的土壤。

Scarlet Grevillea is shade-hardy but still prefers full sun and well-drained soils.

植物趣聞 ANECDOTE ON PLANTS

昆士蘭銀樺和銀樺 Scarlet Grevillea and *Grevillea robusta* (Silk Oak)：

昆士蘭銀樺和銀樺皆屬於山龍眼科，兩者均為公園及花園的常見觀賞品種。由於它們具有類似的特徵，總是被混淆。要區分兩者，可從葉形及花色入手。首先，昆士蘭銀樺的葉片為羽狀深裂，而銀樺的葉片則為二回羽狀深裂。此外，昆士蘭銀樺的花色為紅色，銀樺的花色則為橙色。

Scarlet Grevillea and Silk Oak are members of Proteaceae. They are confused with each other because they share high similarities. We can distinguish the trees by looking at the leaf shape and flower colour. First, the leaves of Scarlet Grevillea are pinnatiparted, whereas those of Silk Oak are bipinnatiparted. Moreover, the blossoms of Scarlet Grevillea are scarlet, while those of Silk Oak are orange.

紫薇

Common Crape Myrtle, Carpe Myrtle | *Lagerstroemia indica* L.

相片拍攝地點：富東邨、樂群街公園、青衣公園
Tree Location: Fu Tung Estate, Lok Kwan Street Park, Tsing Yi Park

名字由來 MEANINGS OF NAME

為紀念瑞典業餘植物學家—馬格努斯‧拉格斯特倫（1691-1759），屬名特此取其名，並為紫薇冠名為 *Lagerstroemia*。種加詞 *indica* 意指印度。

The generic name *Lagerstroemia* commemorates Magnus von Lagerstrom (1696-1759) a Swedish amateur botanist. The species epithet *indica* means from India.

應用 APPLICATION

紫薇作為理想的觀賞樹種，不得不提的是它那雍容華貴的花朵，令人不自覺沉醉其中。百花在夏日盛放時，為城市注入生機，讓途人仿似置身於美輪美奐的風景之中。此樹種的可塑性高，既能以樹的形式種植，又可以種植作灌木叢。

本地分佈狀態 DISTRIBUTIONS	外來物種 Exotic Species
原產地 ORIGIN	印度東北部至南中國，以及緬甸、尼泊爾、越南等東南亞國家。 From Northeastern India to South China, also Southeast Asian countries, such as Myanmar, Nepal and Vietnam.
生長習性 GROWING HABIT	灌木或小落葉喬木。高度可達 7 米。 Shrub or small deciduous tree. Up to 7 m tall.

1	2	3	4	5	6	7	8	9	10	11	12

花果期
月份

花期：本港六月至八月。果期：本港七月至十月。
Flowering period: June to August in Hong Kong. Fruiting period: July to October in Hong Kong.

紫薇具有極高的藥用價值。在紫薇的原生地，它的根部常用於清熱利尿。其葉片可以加工成瀉藥或治療熱毒、大出血和祛濕。此外，其木材堅硬，不易腐爛，故成為廣受歡迎的木材，常用於家具製作。

Common Crape Myrtle is an ideal ornamental tree attributed to its wonderful blossoms. During the flowering seasons, its crown is festooned by delicate purple or pink blossoms, which immerses the city into an alternative aura of vitality and create a unique and picturesque urban landscape.

The tree also serves versatile medicinal functions. In its native range, the roots are always used for expelling heat and promoting diuresis. The leaves can be processed into purgatives or treat heat-toxin, hemorrhage and attenuate dampness. Moreover, its wood is hard and rot-resistant, thereby rendering its popularity to furniture making.

辨認特徵 TRAITS FOR IDENTIFICATION

① 樹幹 TRUNK	② 樹皮 BARK	③ 葉 LEAVES
④ 花 FLOWERS	⑤ 果 FRUITS	

① 紫薇的樹幹。

Trunk of *Lagerstroemia indica* L.

② 樹皮光滑，呈黃棕色。小枝細長，具棱角或狹翅，漸變無毛。

Bark smooth, yellowish brown. Branchlets slender, angular or narrowly winged, glabrescent.

③ 葉片互生至近對生，無柄或具極短的葉柄。葉片紙質至革質，橢圓形、寬長橢圓形或倒卵形，基部寬楔形或近圓形，頂端銳形或鈍形，有時微缺，無毛或沿葉背葉脈被微柔毛。

Simple leaves alternate to subopposite, sessile or with very short petioles. Blade chartaceous to coriaceous, elliptic, broadly oblong or obovate, base broadly cuneate or suborbicular, apex acute or obtuse, sometimes emarginate, glabrous or puberulent along veins abaxially.

④ 圓錐花序頂生，被微柔毛，花朵密集。雌雄同體，花萼合生，具萼筒，花瓣爪形，具皺摺，呈紫色或粉紅色，雄蕊眾多，二形，花萼上的 6 條雄蕊遠比其他雄蕊長。

Panicles terminal, puberulous, dense-flowered. Flowers hermaphroditic, synsepalous, with hypanthium, petals clawed, wrinkled, purple or pink, stamens many, dimorphic, with outer 6 much longer than others.

⑤ 蒴果呈紫黑色，橢圓形，室背開裂。種子具翅。

Capsules purple-balck, ellipsoidal, loculicidal. Seed winged.

生命力 VITALITY

紫薇對乾旱土壤具良好耐受性，但較偏愛肥沃、濕潤、含鈣質或酸性土壤。

Common Crape Myrtle is well tolerant of arid soils and prefers fertile, moist, calcium-containing or acidic soils.

植物趣聞 ANECDOTE ON PLANTS

紫薇屬 *Lagerstroemia*：

紫薇屬的花朵瑰麗。大花紫薇和紫薇皆被廣泛種植在香港的公園和街道上。兩者的區別在於大花紫薇的葉片及樹木體積較紫薇大。

紫薇以花色眾多而聞名，從紅色至紫色，千變萬化。白花的紫薇為栽培種，中文名稱為白花紫薇，學名為 *L. indica* 'Alba'。

Lagerstroemia is a genus of extremely graceful flowering plants. In Hong Kong, *L. speciosa* (Queen Crape Myrtle) and Common Crape Myrtle are the most dominant in parks and streets. The main differences between the two species are that Queen Crape Myrtle showing notably larger leaves and tree sizes.

Common Crape Myrtle is noted with versatile floral colours, ranging from red to purple. The tree with white flowers is a cultivar. The English name is White Carpe Myrtle and the scientific name is *L. indica* 'Alba'.

土沉香 又稱：牙香樹、白木香、女兒香

Incense Tree | *Aquilaria sinensis* (Lour.) Spreng.

相片拍攝地點：獅子會自然教育中心、香港公園、香港動植物公園
Tree Location: Lions Nature Education Centre, Hong Kong Park, Hong Kong Zoological and Botanical Gardens

名字由來 MEANINGS OF NAME

羅香林教授所著的《一八四二年以前之香港及其對外交通》提到，為了讓沉香木及其他副產品如線香，出口到東南亞，土沉香於宋代在香港大量種植。出口沉香木的港口被命名為「香的港口」，象徵沉香木散發出的芳香氣味，該港口進一步更名為「香港」。

Mentioned in the book *Hong Kong and its External Communication before 1842* written by Professor Lo Hsiang-lin in 1963, Incense Tree used to be heavily planted in Hong Kong during the Sung Dynasty for supporting the export of agarwood and other side products such as incense to Southeast Asia. The harbour crammed with the exporting activities was named as "Fragrant Harbour", representing the aromatic smell emitted from the wood. It has been soon renamed as "Hong Kong", a homophone of Fragrant Harbour in Chinese.

本地分佈狀態 DISTRIBUTIONS	原生物種 Native species
原產地 ORIGIN	中國華南地區，包括海南、廣東、廣西和香港。 Southeast China, e.g. Hainan, Guangdong, Guangxi and Hong Kong.
生長習性 GROWING HABIT	常綠喬木。高度可達 15 米。 Evergreen tree. Up to 15 m tall.

花果期 月份

1	2	3	4	5	6	7	8	9	10	11	12

花期：本港三月至五月。果期：本港九月至十月。
Flowering period: March to May in Hong Kong. Fruiting period: September to October in Hong Kong.

　　沉香是一種奢侈品，燃燒時散發出香味。著名品牌 Tod Ford 出品，廣受歡迎的香水正是從土沉香提取及加工而成。除香水外，土沉香還可加工成雕塑或木珠。此外，土沉香更是一種名貴中藥，用以根治腹脹、胃寒、腎虛及哮喘。

Agarwood is an extravaganco with its emblematic fragrance when burnt. The perfumes of the famous brand Tod Ford is produced from extracting the aromatic resin from the wood. Moreover, it is a precious Chinese medicine to treat abdominal distension, stomach cold, kidney deficiency and asthma.

辨認特徵 TRAITS FOR IDENTIFICATION

① 樹幹 TRUNK	② 樹皮 BARK	③ 葉 LEAVES
④ 花 FLOWERS	⑤ 果 FRUITS	

① 土沉香的樹幹。枝條被細柔毛、平滑並呈圓柱狀。

Trunk of *Aquilaria sinensis* (Lour.) Spreng. Branchlets puberulous, smooth and terete.

② 樹皮呈深灰色。

Bark of trunk grayish white.

③ 葉片互生，革質，近卵形，倒卵形至橢圓形，基部寬楔形，頂端短漸尖，兩面皆無毛。葉面呈暗綠色或紫綠色，具光澤，葉背呈淺綠色。側脈纖細，密集，近平行。

Simple leaves alternate. Blade leathery, subovoate, obovate to elliptic, base broadly cuneate, apex short acuminate, glabrous on both surfaces, adaxially dark green, glossy, abaxially pale green. Lateral veins slender, dense, subparallel, indistinct.

④ 花朵細小，頂生。萼筒淺鐘狀，5 裂。花瓣 10 片，鱗片狀，着生於萼筒喉，被柔毛。

Umbels terminal, densely pubescent. Calyx tube shallowly campanulate, 5-lobed, petals 10, scale-like, inserted at the throat of calyx-tube, hairy.

⑤ 蒴果，卵狀球形，呈綠色，基部漸尖，頂端突尖，表面被黃棕色絨毛。成熟時開裂成兩半，2粒種子下垂，具絲狀結構附屬體。種子卵球形，呈深褐色，基部有尾狀附屬體。

Capsules, obovoid, green, base tapering, apex apiculate, surface covered by yellowish brown hairs. Splitting into halves when mature, 2 seeds pendulous with the attachment of filiform structure. Seeds ovoid, dark brown, with tail-shaped appendage at the base.

生態 ECOLOGY

不同研究指出夜蛾科和螟蛾科均是土沉香的重要傳粉媒介。此外，大黃蜂（通常是 *Vespa* sp.）對垂吊在果實下方的附屬體心醉神迷。土沉香的獎勵機制吸引大黃蜂們在飛行過程中協助傳播種子。

According to different studies, both Noctuidae and Pyralidae are important pollinators of *Aquilaria sinensis*. In addition, *Vespa* sp. are fascinated by appendages that hang beneath the fruit. Incense Tree's reward system attracts *Vespa* sp. to help spread its seeds during flight.

植物趣聞 ANECDOTE ON PLANTS

香港稀有及珍貴植物 Rare and Precious Plants of Hong Kong：

土沉香在世界各地皆受到高度保護。它在國際自然保護聯盟瀕危物種紅色名錄中被列為「易危」，在《中國植物紅皮書》中被列為「易危」，同樣被《香港稀有及珍貴植物》評為「近危」。目前，根據《保護瀕危動植物物種條例》（香港法例第 586 章），未經漁農自然護理署簽發的許可證，皆被禁止進口、出口或再出口任何瀕危物種。縱使受到各種條例保護，因其珍罕程度極高，沉香木仍面臨着嚴峻的非法採伐問題。

在 2011 至 2019 年期間，香港一共有 1360 棵土沉香被砍伐，當中 998 棵為非法砍伐。此外，走私土沉香的數量驚人，被香港海關查獲的土沉香總價值高達 3970 萬港元。

The wild population of Incense Tree is being highly protected over the world. As a result, the tree is currently graded as Vulnerable in the International Union for Conservation of Nature (IUCN) Red List of Threatened Species and *China Plant Red Data Book*. It is also rated as Near Threatened in *Rare and Precious Plants of Hong Kong*. According to the Protection of Endangered Species of Animals and Plants Ordinance (Cap. 586) in Hong Kong, importing, exporting or re-exporting of the tree without any license issued by the Agriculture, Fisheries and Conservation Department (AFCD) is forbidden. Despite the increase of awareness by the governments to the tree, it is still embroiled in severe illegal logging due to the desperate yearns for its wood.

Despite the increase of awareness by the governments to the tree, it is still embroiled in severe illegal logging due to the desperate yearns for its wood. From 2011 to 2019, a total of 1360 Agarwood trees were felled in Hong Kong, of which 998 were illegally felled. At the same time, a startling number of smuggled agarwood, with HKD 39.7 million worth of Incense wood, were seized by Hong Kong customs.

紅千層 又稱：瓶子刷樹、紅瓶子刷樹

Stiff Bottle-brush, Rigid Bottle-brush | *Callistemon rigidus* R. Br.

相片拍攝地點：九龍公園
Tree Location: Kowloon Park

名字由來 MEANINGS OF NAME

屬名是由 *callis* 和 *stemon* 兩個希臘詞組成的混合字，分別指「美麗」和「雄蕊」，暗指其雄蕊是花朵中最迷人的部分。由於其枝條堅挺直立，花朵形似瓶刷，故又名「紅瓶刷子樹」。

The generic name is a blend of *callis* and *stemon*, the Greek words that refer to "beautiful" and "stamen", together alluding to its stamens as the most glamorous part of the flowers. Since the tree shows firm and erect branches and bottlebrush like flowers, it is also named as "Stiff Bottle-brush".

應用 APPLICATION

紅千層的花朵如火焰般艷紅奪目，故被視為絕佳的觀賞樹種，常被種植於公園和花園內。盛夏時，花序從小枝頂端伸出，整個樹冠被連綿不斷的花朵覆蓋；花落時，朵朵紅花從樹上傾瀉而下，為地面鋪上一層紅地毯。

本地分佈狀態 DISTRIBUTIONS	外來物種 Exotic species
原產地 ORIGIN	原生於澳洲。目前亦被廣泛栽培於廣東及廣西。 Native to Australia. It is now broadly cultivated in Guangdong and Guangxi.
生長習性 GROWING HABIT	常綠小喬木。高度可達 5 米。 Small evergreen tree. Up to 5 m tall.

1	2	3	4	5	6	7	8	9	10	11	12

花果期月份

花期：本港六月至八月。果期：本港十二月至二月。
Flowering period: June to August in Hong Kong. Fruiting period: December to February in Hong Kong.

可從紅千層的葉片中提取精油並加工成化妝品、肥皂和洗滌劑。其枝葉可用作傳統中藥，具有化痰消腫並紓緩感冒、類風濕性關節炎及濕疹之效。

By virtue of flamboyant scarlet flowers, the tree is considered as a gorgeous ornamental component for parks and gardens. In summer, the inflorescences protrude from the apex of branchlets, blanket the crown in endless blooms, rain down and bury the ground. It lays a red blanket on the ground when the flowers fall from the trees.

The essential oil can be extracted from its leaves and processed into cosmetics, soaps and detergents. The branches and leaves can also be used for traditional Chinese medicines, functionally effective to relieve phlegm and swelling, and attenuate common cold, rheumatic arthralgia and eczema.

辨認特徵 TRAITS FOR IDENTIFICATION

① 樹幹 TRUNK	② 樹皮 BARK	③ 葉 LEAVES
④ 花 FLOWERS	⑤ 果 FRUITS	

① 紅千層的樹幹。

Trunk of *Callistemon rigidus* R. Br.

② 樹皮粗糙，棕灰色，片狀剝落。小枝具棱，幼時被長絲毛，光滑無毛。

Bark coarse, brownish grey, peeling off in flakes. Branchlets angulate, covered with long silky hairs when young, glabrescent.

③ 單葉互生。葉片革質，密被油腺體，光滑無毛，線狀，頂端銳形，兩面中脈凸起。

Simple leaves alternate. Blade leathery, covered with dense oil glands, glabrescent, linear, apex acute, midvein elevated on both surfaces.

④ 雌雄同體。穗狀花序頂生，嫩葉萌發時轉為腋生，幼時花萼和花軸被短柔毛。雄蕊眾多，較長，花絲呈朱紅色，花藥呈黃色。

Hermaphroditic. Spikes terminal, axillary-like when new leaves emerge, calyx and rachis pubescent when young. Stamens many, long, filaments scarlet, anthers yellow.

⑤ 蒴果，木質，半球形，頂端截形，3 片裂開，開裂時爆開，宿存。

Capsules woody, semi-globose, apex truncate, valves 3, split when dehiscent, persistent.

生命力 VITALITY

紅千層對乾旱及陰暗環境具高耐受性，並偏愛微酸、潮濕且排水良好的土壤。

Stiff Bottle-brush is highly tolerant to drought and shade. It prefers slightly acidic, moist and well-drained soils.

植物趣聞 ANECDOTE ON PLANTS

紅千層與串錢柳 Stiff Bottle-brush and *Callistemon viminalis* (Tall Bottle-brush)：

紅千層與串錢柳皆屬於紅千層屬。兩者的外形相似，難以辨認，一般紅千層被視為串錢柳的替代品並種植於公園內。樹形是區分兩者的關鍵。串錢柳亦有 Tall Bottle-brush 之稱，顧名思義，串錢柳較紅千層高，最高可達 10 米。此外，串錢柳的花朵和枝條不如紅千層般粗壯結實，而是優雅地向下垂，隨風搖曳。

Both of the trees are the members of *Callistemon*. In respect of their indistinguishable appearances, Stiff Bottle-brush is always regarded as a visual alternative of Tall Bottle-brush for parks. Tree form is rather an unmistakably diagnosed key to distinguish the trees. As how it is named, Tall Bottle-brush is relatively taller, with an attainable height to 10 metres. Likewise, instead of standing stout, its flowers and branches are pendulous and gracefully hand down to the ground.

水翁 又稱：水翁蒲桃、水榕

Lidded Cleistocalyx, Water Banyan | *Cleistocalyx nervosum* (DC.) Kosterm.

相片拍攝地點：獅子會自然教育中心、雷公田
Tree Location: Lions Nature Education Centre, Lui Kung Tin

名字由來 MEANINGS OF NAME

種加詞 *nervosum* 意指此物種的凸出葉脈。

The specific epithet *nervosum* refers to its prominent leaf veins.

應用 APPLICATION

水翁具藥用價值，幾乎整株植物皆可入藥，包括葉片、花朵、根部和樹皮。它的樹皮具清熱、利濕、驅蟲之效；葉片可消滯和紓緩瘙癢；花朵是廣東知名涼茶「廿四味」和「五花茶」的重要原料。除醫藥用途外，水翁果實多汁，可生食或加工成酒。

由於水翁具豐富的生態價值，故是一種具潛力的本地樹種，可用於香港再造林。

本地分佈狀態 DISTRIBUTIONS	原生物種 Native species
原產地 ORIGIN	廣東、廣西及雲南。同時亦分佈在中南半島、印度、馬來西亞、印尼及大洋洲等地。 Guangdong, Guangxi and Yunnan. It is also distributed in Indochina, India, Malaysia, Indonesia and Oceania.
生長習性 GROWING HABIT	常綠喬木。高度可達 15 米。 Evergreen tree. Up to 15 m tall.

花果期 月份

1	2	3	4	5	6	7	8	9	10	11	12

花期：本港五月至六月。果期：本港八月至九月。
Flowering period: May to June in Hong Kong. Fruiting period: August to September in Hong Kong.

The tree is exploited for traditional Chinese medicines due to its versatility of medicinal probabilities. The bark is always used for relieving heat, dampness and parasites; the leaves can attenuate stagnation and itching; the flower buds are one of the primary ingredients of "24 herbal tea" and "five flower tea", which are the prominent herbal teas in Guangdong. The fruits are juicy and wholesome, thereby always eaten fresh or processed into wines.

By reason of its multiple ecological benefits, the tree serves as a potential native species for reforestation work in Hong Kong.

| ① 樹幹 TRUNK | ② 樹皮 BARK | ③ 葉 LEAVES |
| ④ 花 FLOWERS | ⑤ 果 FRUITS | |

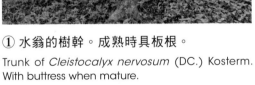

① 水翁的樹幹。成熟時具板根。

Trunk of *Cleistocalyx nervosum* (DC.) Kosterm. With buttress when mature.

② 樹皮頗厚，灰棕色，塊狀剝落，具縱裂痕。

Bark rather thick, greyish brown, massive exfoliation, with longitudinal cracks.

③ 葉片對生，葉柄腹面具凹槽，葉片長橢圓形至橢圓形，稍革質，基部寬楔形至近圓形，頂端銳形至漸尖。葉背中脈隆起，側脈於中脈45°至60°開角斜向上，邊脈明顯。

Simple leaves opposite. Petioles grooved adaxially. Blade glabrous, oblong to elliptic, slightly leathery, base broadly cuneate to slightly rounded, apex acute to acuminate, entire, midvein raised abaxially, pointing at an angle of 45°-60° from midvein, reticulate and intramarginal veins obvious.

163

④ 聚傘圓錐花序生於無葉枝條。花雙性，常 2-3 朵簇生，無柄。帽狀體頂端具短喙，被絲托半球形。雄蕊眾多，呈白色。

Hermaphroditic. Thyrses axillary, on leafless branches. Flowers bisexual, always 2-3 in clusters, sessile. Hypanthium semiglobose, calyptra apex beaked, stamens many, white.

⑤ 漿果寬卵球形，成熟時轉為紫黑色。

Berries broadly ovoid, turning purple black at maturity.

生態 ECOLOGY

　　水翁的花朵具獨特芳香，可以吸引昆蟲。常見蝴蝶品種如青鳳蝶、相思帶蛺蝶和橙端粉蝶經常造訪水翁。此外，它的漿果色彩艷麗，深受鳥類喜愛。

The blossoms of Lidded Cleistocalyx are distinctively fragrant and extremely attractive to insects. Some common butterfly species like *Graphium sarpedon* (Common Bluebottle), *Athyma nefte* (Colour Sergeant), and *Hebomoia glaucippe* (Great Orange-tip) are the common butterfies that always visit the flowers. Furthermore, its berries are colourful, serving as a dellcacy to birds.

生命力 VITALITY

　　水翁常見於溪流和沼澤。它對澇漬和多風的環境具高耐受性，但不能忍受乾旱環境。

Lidded Cleistocalyx is mainly distributed along streams and swamps. It is highly tolerant to waterlogging and windy environments, but cannot tolerate drought.

毛葉桉

Cadaga | *Corymbia torelliana* (F. Muell.) K. D. Hill & L. A. S. Johnson

相片拍攝地點：荔枝角公園、鳳德公園
Tree Location: Lai Chi Kok Park, Fung Tak Park

名字由來 MEANINGS OF NAME

屬名 *Corymbia* 是指其花朵簇生在傘形花序中。種加詞 *torelliana* 是為了紀念意大利參議員——路易吉·托雷利伯爵（1810-1887），他提議種植桉屬樹種來降低蓬蒂內沼澤的濕度，以減低感染瘧疾的風險。

The generic name *Corymbia* refers to its blossoms clustered in an umbel. The specific epithet *torelliana* was named after Count Luigi Torelli (1810-1887), a member of the Italian senate who proposed the application of eucalyptus for drying the Pontine Marshes for the sake of minimizing the risk of malaria.

本地分佈狀態 DISTRIBUTIONS	外來物種 Exotic species
原產地 ORIGIN	原生於澳洲北昆士蘭。 North Queensland in Australia.
生長習性 GROWING HABIT	大型常綠喬木。高度可達 30 米。 Large evergreen tree. Up to 30 m tall.

1	2	3	4	5	6	7	8	9	10	11	12

花果期月份

花期：本港十月至十一月。果期：不詳。
Flowering period: October to November in Hong Kong. Fruiting period: Unknown.

　　毛葉桉的葉片具藥效，可治療由細菌感染引起的喉嚨痛和呼吸困難；葉片萃取物還可治療肺病。由於毛葉桉擁有高大和健壯的樹幹，許多城市的防風林皆會種植此品種，如美國的佛羅里達州。

The leaves can treat sore throat and unpleasant breathing triggered by bacterial infection; the leaf extracts can treat lung disease. The towering height and robust trunk render the tree excellent windfirm and thus has been exploited for windbreaks in many countries such as Florida of the United States.

① 樹幹 TRUNK	② 樹皮 BARK	③ 葉 LEAVES
④ 花 FLOWERS	⑤ 果 FRUITS	

① 毛葉桉的樹幹。

Trunk of *Corymbia torelliana* (F. Muell.) K. D. Hill & L. A. S. Johnson.

② 樹皮光滑，呈灰綠色至白色，粗糙、暗褐色樹皮宿存於樹幹基部。小枝圓柱狀，初時密被絨毛，木質化後無毛。

Bark greyish green to white, smooth, exfoliating but persistent at trunk base, coarse, dark brown. Branchlets terete, hirsute when young, glabrescent.

③ 單葉互生，幼葉被紅色絨毛，葉片成熟後轉為薄革質，卵形，基部圓形，頂端銳形至漸尖，葉緣全緣，葉背被短柔毛。

Simple leaves alternate. Young leaves red pubescent. Adult leaves thinly leathery, ovate, base round, apex acute to acuminate, margin entire, abaxially pubescent. Petioles hirsute. Young leaves and petioles red pubescent.

④ 聚傘圓錐花序頂生或腋生，每個繖形花序常具 3-7 朵花。花托筒半球形，帽狀體呈綠色，圓形至圓錐形至稍具喙。雄蕊眾多，呈白色。

Thyrse terminal or axillary, each umbel always 3-7 flowered. Hypanthium semiglobose, calyptra green, rounded or conical to slightly beaked. Stamens many, white.

⑤ 蒴果球形，甕形，花盤凹入，3 瓣開裂。

Capsules globose, urn-shaped, disk contracted, valves 3, included in hypanthium.

生態 ECOLOGY

毛葉桉的種子主要依靠重力和蜜蜂傳播。當果實成熟時，大部分種子可以從果瓣開口中掉落在附近的地面上。雖然重力是一種較省力的傳播方式，但限制了種子分散在一定範圍內，故加劇種內競爭。

在「物競天擇，適者生存」的壓力下，毛葉桉與蜜蜂抱持着互利共生的關係。毛葉桉的果實會在每個果瓣邊緣分泌黏性樹脂，這有機會黏緊部分種子，以致種子未能按重力掉下。樹脂對蜜蜂具高度吸引力。蜜蜂會利用樹脂築巢，過程中有機會將含樹脂的種子運送到巢中。由於蜜蜂會把種子遠距傳播，若能把種子帶到陽光充足的地方，便能提高種子存活率。

The seeds of Cadaga are primary driven by gravity and bees. When the fruit is mature, it pops and most of the seeds fly out from the valves and drop on the ground nearby. Although gravity is an energy-saving dispersal method, it limits the distance of seed dispersal and hence intensifies the intraspecific competition.

Under the selective pressure, Cadaga has engaged in a mutualistic interaction with bees. The tree secretes sticky resin droplets at the edge of every valve, for the sake of clinging the seeds away from the gravitational force. The resins are attractive to bees and the bees will use the resin to build nests. The bee may transport the resin-containing seeds to their nests through the process. The long dispersal is thought to increase the survival rate for higher probability to germinate in a place with sufficient sunlight.

生命力 VITALITY

毛葉桉生長迅速，對乾旱、高溫、寒冷、風和貧瘠土壤具高耐受性。

Cadaga is fast-growing and highly tolerant to drought, heat, cold, wind and barren soils.

赤桉

River Red Gum, Murray Red Gum | *Eucalyptus camaldulensis* Dehnh.

相片拍攝地點：百福田心遊樂場
Tree Location: Pak Fuk Tin Sum Playground

名字由來 MEANINGS OF NAME

屬名 *Eucalyptus* 源自希臘語，意譯為被遮蓋，暗指此屬植物的花朵完全隱藏在帽狀體中。赤桉的首個標本製作及相關描述源於卡馬爾多利，故被冠名為 *camaldulensis*。赤桉具紅色的木材和沿河岸生長的特性，故亦有「河流紅膠樹」之稱。

The generic name *Eucalyptus* is derived from the Greek words (well-covered), describing the flowers in this genus are well-concealed in a calyptra. The specific epithet *camaldulensis* refers to Camaldoli in Italy, where the first specimen of the tree was given a botanical description. The tree is also named as "River Red Gum" in respect of its red timber and the propensity to grow along riverbanks.

本地分佈狀態 DISTRIBUTIONS	外來物種 Exotic species
原產地 ORIGIN	澳洲。 Australia.
生長習性 GROWING HABIT	常綠喬木。高度可達 25 米。 Evergreen tree. Up to 25 m tall.

花果期 月份	1	2	3	4	5	6	7	8	9	10	11	12

花期：本港十二月至八月。果期：本港九月至十月。
Flowering period: December to August in Hong Kong. Fruiting period: September to October in Hong Kong.

　　赤桉生長速度快，心材紅潤，抗腐蝕能力強，且可彎性高，故被廣泛種植以取其木材。其木材常用於製作枕木、木樁、家具、瓷磚和工具把手。其枝葉具有清熱解毒、抗過敏等藥用價值。

River Red Gum is broadly planted for lumber by virtue of its fast-growing, beautiful red heartwood, notable resistance to corrosion, and high bendability. The wood is highly demanded for making sleepers, woodpiles, furniture, tiles and tool handles. The branches and leaves are valued for their wonderful medicinal effects for expelling heat, toxins and allergy.

辨認特徵 TRAITS FOR IDENTIFICATION

① 樹幹 TRUNK	② 樹皮 BARK	③ 葉 LEAVES
④ 果 FRUITS		

① 赤桉的樹幹。

Trunk of *Eucalyptus camaldulensis* Dehnh.

② 全株樹皮光滑，灰白色至暗灰色，片狀剝落。

Bark smooth throughout, greyish white to dark grey, exfoliating in flakes.

③ 單葉互生。葉片成熟後呈薄革質，暗綠色，窄披針形，稍鐮形，頂端漸尖，兩面具黑色小腺點，密被網狀紋。

Simple leaves alternate. Mature blade thinly leathery, dull green, narrowly lanceolate, slightly falcate, apex acuminate, both surfaces with small black glands, densely reticulate.

④ 蒴果，近球形，果瓣 3-5 片，果實着生於萼管。種子成熟時由黃色轉為淡褐色。

Capsules subglobose, valves 3-5, exerted from hypanthium. Seeds yellow to light brown.

註：本樹另有花，雌雄同體。繖形花序，腋生，花朵 5-11 朵。花蕾卵形，帽狀體具喙，呈綠色。雄蕊眾多，呈白色。

Remarks: Flowers Hermaphroditic. Umbels axillary, 5-11-flowered. Flower buds ovate, calyptra beaked, green. Stamens many, white.

生命力 VITALITY

　　赤桉對乾旱的耐受性高，在降水不足的環境下仍能生存。然而，它偏愛陽光充沛和潮濕的土壤。

It is highly tolerant to drought and can survive where the precipitation is low. Yet, it still prefers sunny weather and moist soils.

植物趣聞 ANECDOTE ON PLANTS

如何辨認桉屬？How to identify *Eucalyptus*？：

　　桉屬已廣泛引入至香港作林地復育。桉屬種類多不勝數，其樹高及模糊的形態差異導致難以精確鑑別。要辨認它們，可從其樹皮、葉片、帽狀體、蒴果及種子顏色入手。舉例來説，檸檬桉的樹皮光滑，大葉桉的樹皮則具宿存。大葉桉的葉片呈卵狀披針形，赤桉的葉片則呈狹披針形。毛葉桉的帽狀體呈圓形，窿緣桉的帽狀體則呈圓錐狀。尾葉桉的蒴果果瓣凹入，赤桉的蒴果果瓣凸出。赤桉的種子呈黃色，細葉桉的種子則呈黑色。

Eucalyptus has been widely introduced in Hong Kong for rehabilitation. There are numerous species of *Eucalyptus* (and *Corymbia*), and their lofty height and obscure morphological differences make it difficult to distinguish the different species. The differences between their bark, leaves, calyptras, capsules and seed colour are some keys to differentiate them. For example, bark of Lemon-scented Gum (*Corymbia citriodora*) is smooth throughout while that of Swamp Mahogany (*E. robusta*) is persistent, leaves of *Swamp Mahogany* are ovate-lanceolate while that of River Red Gum are narrowly-lanceolate, calyptras of Cadaga (*C. torelliana*) is rounded while those of Queensland Peppermint (*E. exserta*) is conic, valves of capsules of Timor White Gum (*E. urophylla*) are included while those of River Red Gum are exserted, and seed colours of River Red Gum is yellow while that of Forest Gray Gum (*E. tereticornis*) is black.

窿緣桉

Queensland Peppermint | *Eucalyptus exserta* F. Muell.

相片拍攝地點：維多利亞公園對出、沙田公園
Tree Location: Next to Victoria Park, Sha Tin Park

名字由來 MEANINGS OF NAME

窿緣桉的果瓣明顯凸出，故被冠名為 *exserta*（中文意譯為凸出）。由於首次收集到窿緣桉的地點為昆士蘭（澳洲），而且當碾碎其葉片時會散出陣陣清爽的薄荷氣味，故亦有 Queensland Peppermint（中文意譯為昆士蘭胡椒薄荷）此俗名。

The specific epithet *exserta* describes its fruits with strongly exserted valves. The common name Queensland Peppermint refers to the type specimen from Queensland and leaves smelling like mint when crushed.

本地分佈狀態 DISTRIBUTIONS	外來物種 Exotic species
原產地 ORIGIN	原生於澳洲。廣泛栽培於華南地區，包括福建、廣東、廣西和海南。 Native to Australia. Widely cultivated in Southern China, including Fujian, Guangdong, Guangxi and Hainan.
生長習性 GROWING HABIT	常綠喬木。高度可達 18 米。 Evergreen tree. Up to 18 m tall.

1	2	3	4	5	6	7	8	9	10	11	12

花果期月份

花期：本港五月至九月。果期：本港七月至十一月。
Flowering period: May to September in Hong Kong. Fruiting period: July to November in Hong Kong.

窿緣桉的樹皮粗糙、堅硬且耐用，並可加工成屋樑、柵欄及家具。其葉片油量豐富，而且在民間治療濕疹和疥瘡的偏方中擔任重要材料。其萃取物能對抗一種危害糧食存備的害蟲——赤擬穀盜。

桉屬生長速度快，且對強風和乾旱對高耐受性，加上能適應亞濕潤氣候至熱帶氣候，故在氣候差異大的澳洲中成為優勢種。基於以上原因，桉屬一直被視為先鋒樹種，用作防風林及林地復育，以減少財產損失。

The timber is rough, hard and durable, and is always processed into house framing, fencing and furniture. The profusely oily leaves are the primary ingredient of Chinese folk medicine to treat skin eczema and scabies. The leaf extracts are antagonistic to *Tribolium castaneum* (Red flour beetle), a rampant pest which is always found in stored food products.

Eucalyptus is a fast-growing species with high tolerance to strong winds and droughts. It can also adapt to sub-humid to tropical climates, making it the dominant species in Australia. Based on these reasons, *Eucalyptus* are always regarded as superior windbreak and pioneer trees used for windbreaks and woodland restoration to reduce property losses.

辨認特徵 TRAITS FOR IDENTIFICATION

① 樹幹 TRUNK	② 樹皮 BARK	③ 葉 LEAVES
④ 花 FLOWERS	⑤ 果 FRUITS	

① 窿緣桉的樹幹。幼枝常下垂。

Trunk of *Eucalyptus exserta* F. Muell.

② 樹皮灰棕色，粗糙，堅硬，具縫裂、宿存。

Bark greyish brown, rough, hard, fissured, persistent. Young branchlets always pendulous.

③ 幼葉對生，葉柄較短。葉成熟後轉為互生，葉柄纖細，葉片狹披針形，略帶光澤或暗啞，兩面同色。

Juvenile leaves opposite, petiole short. Adult leaves alternate, petiole slender, blade narrowly lanceolate, slightly glossy or dull, concolourous.

④ 腋生，每個傘形花序具 3-8 個花蕾。花芽卵球形，帽狀體長圓錐形，頂端漸尖，呈淡黃色或乳白色，被絲托半球形。雄蕊呈白色，直立，眾多。

Axillary, flower buds 3-8 per umbel. Flower buds ovoid, calyptra long conic, apex acuminate, yellow or cream, hypanthium semiglobose. Stamens white, stand upright, many.

⑤ 蒴果近球形，花盤凸起，果瓣 4 片，明顯凸出。

Capsules subglobose, floral disc raised, valves 4, strongly exserted.

生命力 VITALITY

　　窿緣桉的棲息地範圍涵蓋亞濕潤氣候至熱帶氣候。其對乾旱具高耐受性，但偏愛排水良好的土壤。

The range of Queensland Peppermint habitat ranges from sub-humid to tropical climates. It shows exceptionally tolerance to drought, but prefers well-drained soils.

細葉桉

Forest Gray Gum, Forest Red Gum | *Eucalyptus tereticornis* Sm.

相片拍攝地點：港鐵火炭站
Tree Location: Fo Tan MTR Station

名字由來 MEANINGS OF NAME

種加詞 *tereticornis* 意為其長而像角的帽狀體。

The specific epithet *tereticornis* refers to its long and horn-like calyptra.

應用 APPLICATION

細葉桉的葉片含豐富油脂，常加工作香料、殺幼蟲油及藥油。其木材耐用、細膩、密度高、紋路交錯；故常用於製造柴薪、鋪路材料及風車軸承。細葉桉是一種快速生長的樹種，是造林時主要使用的桉樹樹種。在香港南生圍有一個大型細葉桉種植園。

The leaves of it contains rich essential oils. The oils derived are usually processed into fragrance, larvicidal oil and medicinal oil. The wood is durable, fine, dense, with interlocked grain; therefore, it is always used for making fuel wood, paving and windmill bearings. Forest Gray Gum is a fast-growth tree and one of the predominant *Eucalyptus* spp. used in afforestation work. In Hong Kong, there is a large plantation with Forest Gray Gum in Nam Sang Wai.

本地分佈狀態 DISTRIBUTIONS	外來物種 Exotic species
原產地 ORIGIN	新幾內亞和澳洲，包括新南威爾士、昆士蘭和維多利亞。 New Guinea and Australia, including New South Wales, Queensland and Victoria.
生長習性 GROWING HABIT	常綠喬木。高度可達 25 米。 Evergreen tree. Up to 25 m tall.

花果期 月份

1	2	3	4	5	6	7	8	9	10	11	12

花期：本港六月至八月。果期：本港八月至九月。
Flowering period: June to August in Hong Kong. Fruiting period: August to September in Hong Kong.

① 樹幹 TRUNK	② 樹皮 BARK	③ 葉 LEAVES
④ 花 FLOWERS	⑤ 果 FRUITS	

① 細葉桉的樹幹。

Trunk of *Eucalyptus tereticornis* Sm.

② 樹皮光滑，呈灰白色至棕褐色，不規則片狀或長條狀剝落，常在樹幹基部宿存，粗糙，呈黑色。

Bark greyish white to brown, smooth, exfoliating in irregular flakes or long strips, persistent at trunk base, coarse, black.

③ 葉片互生，初時葉片卵形至寬披針形，成熟時轉為窄披針形和稍扭曲，兩面皆為亮綠色、無毛。

Simple leaves alternate. Blade of juvenile tree ovate to broadly lanceolate. Blade of adult tree glabrous, narrowly lanceolate, slightly twisted at maturity, both surfaces bright green, glabrous.

④ 花朵通常 5-8 朵簇生成傘形花序，腋生。帽狀體圓錐形至長圓錐形，被絲托半球形。雄蕊眾多，呈白色。

Flowers always 5-8 clustered in an umbel inflorescence, axillary. Flowers buds long ovoid, calyptra conical to long conical, hypanthium hemispheric. Stamens many, white, protruded.

⑤ 蒴果近球形至卵形，4-5 裂，花盤凸起，具突出的邊緣。成熟時由綠色轉為暗褐色。

Capsules subglobose to ovoid, floral disk raised, with prominent rim, valves 4-5, strongly exserted. From green to dark brown when mature.

生態 ECOLOGY

　　桉屬的葉片（如細葉桉、大葉桉、赤桉）是樹熊的主要糧食。葉片的油脂對大多數動物來説有毒性，但由於樹熊的消化道中存在微生物，可降解葉片中的毒素，故對樹熊沒有影響，可以安全地從葉片攝取營養。意想不到的是，初生樹熊必須吃掉媽媽的糞便來啟動這個奇妙的消化道防衞系統！

The leaves of *Eucalyptus* spp. (e.g. *E. tereticornis, E. robusta, E. camaldulensis*) serves as the primary food source for koalas. The essential oils from the leaves are pernicious to most animals, however, show no effect on koalas whose digestive tract is armed with special microorganisms that can degrade the toxin. Therefore, koalas can obtain the nutrients from the leaves safely. Surprisingly, a baby koala must eat its mother's faeces to trigger this marvellous defense system of digestive tract!

生命力 VITALITY

　　細葉桉對乾旱和輕微霜凍具耐受性，但其更偏好稍微黏稠的肥沃土壤以茁壯成長。

Forest Gray Gum is resilient to diverse environments such as dry climates and light frosts. but prefers slightly sticky and fertile soils for healthy growth.

黃金香柳

Golden Tea Tree │ *Melaleuca bracteata* 'Revolution Gold'

相片拍攝地點：香港公園、賽馬會德華公園、添馬公園
Tree Location: Hong Kong Park, Jockey Club Tak Wah Park, Tamar Park

名字由來 MEANINGS OF NAME

　　屬名 *Melaleuca* 是由 *melas* 和 *leukos* 組成的希臘語混合字，分別意指「黑色」和「白色」，表示該屬的植物擁有黑色樹幹及白色枝條。種加詞 *bracteata* 意指「具苞片」，形容其苞片於開花時仍附着在花朵上。

　　黃金香柳是野生黃金香柳的栽培品種。根據國際植物命名法規（ICBN）的準則，為強調栽培種的特性，命名栽培種時應在其野生種名後方加上對栽培種的描述，並加上引號。在此，Revolution Gold 形容其栽培種的金綠色葉片遺傳自野生黃金香柳。

本地分佈狀態 DISTRIBUTIONS	**外來物種** Exotic species	
原產地 ORIGIN	**野生黃金香柳原生於澳洲。** The wild type is native to Australia.	
生長習性 GROWING HABIT	**常綠喬木。高度可達 6 米。** Evergreen tree. Up to 6 m tall.	

1	2	3	4	5	6	7	8	9	10	11	12	花果期 月份

花期：不詳。果期：不詳。
Flowering period: Unknown. Fruiting period: Unknown.

The genus name *Melaleuca* is a blend of the Greek words *melas*, meaning black, and *leukos*, meaning white, collectively referring to the members of this genus sharing black trunk and white branchlets. The specific epithet *bracteata* means "with bract", describing its bracts still clinging to the flowers while blooming.

Golden Tea Tree is a cultivar of wild *Melaleuca bracteata*. According to the guideline of International Code of Botanical Nomenclature (ICBN), a cultivar should be named with an additional description in single quotation marks after its wild type's scientific name, primarily for emphasizing its cultivated character. Here, the cultivator epithet 'Revolution Gold' describes its golden green foliage which lineage goes back to the wild Black Tea Tree (*M. bracteata*).

應用 APPLICATION

黃金香柳擁有金燦燦的嫩綠葉片，故被視為極具觀賞價值的樹種。其葉片顏色亮麗獨特，在同是白千層屬的其他植物中獨樹一格，較易辨認。黃金香柳可單獨種植或與其他植物互相配襯，適合種植於所有綠化空間。

黃金香柳的葉片富含茶樹油，碾碎後會釋出陣陣清新怡人的香氣。茶樹油是一種精油，桃金娘科的植物皆含有大量茶樹油。有研究顯示黃金香柳的茶樹油具有顯著的抗氧化及抗菌作用，可見其作藥用和農業用途的潛力。

Golden Tea Tree is regarded as highly ornamental in view of appealing golden-green foliage. The leaf colour is vivid and emblematic enough to be diagnosed from other *Melaleuca* spp. which the leaf shapes are relatively comparable. It can be planted in solitary and with other plants, and is suitable to be planted in all green space.

The leaves are enriched with tea tree oil, a kind of essential oil that is abundant in Myrtaceae, they are aromatic when crushed. The tea tree oil of Golden Tea Tree has excellent antioxidant and antimicrobial effects according to some studies, showing its great potential for pharmaceutical and agricultural purposes.

辨認特徵 TRAITS FOR IDENTIFICATION

① 樹幹 TRUNK	② 樹皮 BARK	③ 葉 LEAVES
④ 花 FLOWERS	⑤ 果 FRUITS	

① 黃金香柳的樹幹。

Trunk of *Melaleuca bracteata* 'Revolution Gold'.

② 樹皮灰色，粗糙，具裂紋。

Bark grey, coarse, fissured.

③ 單葉互生。葉片無毛，呈金綠色，線狀至狹長橢圓形，葉緣全緣，具平行葉脈。

Simple leaves alternate. Blade glabrous, golden-green, linear to narrowly oblong, margin entire, veins parallel.

④ 雌雄同體。穗狀花序，頂生，新葉萌發時花序呈側腋生狀。單生花或 2-3 朵簇生，具香氣，雄蕊眾多，呈白色。

Hermaphroditic. Spikes terminal, lateral axillary-like when new leaves emerge. Flowers solitary or 2-3 clustered, scented, stamens many, white.

⑤ 蒴果，球形，頂端具花萼宿存。

Capsules globular, with persistent calyx at the apex.

生命力 VITALITY

黃金香柳對多變的氣候和土壤具高適應力，但偏愛排水良好的土壤。

Golden Tea Tree adapts well in different climates and soils but grows better in well-drained soils.

備註 Remarks

本樹木學名根據新加坡國家公園管理局網頁：
Scientific name of this tree is based on National Parks Board of Singapore:
https://www.nparks.gov.sg/florafaunaweb/

草莓番石榴

Strawberry Guava | *Psidium cattleyanum* Sabine var. *littorale* (Raddi) Mattos

相片拍攝地點：獅子會自然教育中心
Tree Location: Lions Nature Education Centre

名字由來 MEANINGS OF NAME

為紀念畢生致力於蘭花採集和栽培的英國園藝學家威廉‧卡特利（1788-1835），特此取 *cattleyanum* 作種加詞。此樹種成熟的果實後外果皮顏色猶如鮮紅的草莓，品嚐起來的味道卻像番石榴，故亦有 Strawberry Guava 的俗名（中文意譯為草莓番石榴）。

The specific epithet *cattleyanum* commemorates William Cattley (1788-1835), a British horticulturist who devoted his entire life in plant collection and cultivation of orchids. The tree is also named as "Strawberry Guava" in respect of its mature fruits that tastes like guava and are covered by strawberry-coloured exocarp.

本地分佈狀態 DISTRIBUTIONS	外來物種 Exotic species
原產地 ORIGIN	巴西和烏拉圭。 Brazil and Uruguay.
生長習性 GROWING HABIT	常綠或落葉喬木。高度可達 12 米。 Evergreen or deciduous tree. Up to 12 m tall.

花果期 月份	1	2	3	4	5	6	7	8	9	10	11	12

花期：本港五月至十二月。果期：本港十二月至二月。
Flowering period: May to December in Hong Kong. Fruiting period: December to February in Hong Kong.

　　草莓番石榴美味多汁，可生食，其蘊含豐富抗炎及抗氧化元素，有益健康。此外，亦可加工成果汁、果醬和手工糖。

Strawberry Guava is valued for its delicious and juicy fruits. The fruit can be eaten fresh and is with abundant anti-inflammatory and antioxidant elements and hence good for health. Apart from that, the fruit can be processed into juice, jams and artisanal sweets.

辨認特徵 TRAITS FOR IDENTIFICATION

① 樹幹 TRUNK	② 樹皮 BARK	③ 葉 LEAVES
④ 果 FRUITS		

① 草莓番石榴的樹幹。

Trunk of *Psidium cattleyanum* Sabine var. *littorale* (Raddi) Mattos.

② 樹皮光滑，片狀剝落，呈黃棕色。

Bark smooth, peeling off in flakes, yellowish-brown.

③ 單葉對生，無毛，厚革質，葉背光滑，呈暗綠色，葉片橢圓形，基部楔形，頂端銳形或鈍形，葉緣全緣，基部中脈突出，側脈不明顯。葉柄較短。

Simple leaves opposite, glabrous, thick coriaceous, abaxially dark green and glossy, blade elliptic, base cuneate, apex acute or obtuse, entire, midvein prominent at the base, lateral veins inconspicuous. Petioles short.

④ 漿果球形至卵形，先端具宿存花萼，成熟時由綠色轉為黃色或紫紅色。

Berries globose to ovoid, persistent calyx at the tip, turning green to yellow or purplish-red when mature.

註：本樹另有花，雌雄同體。花腋生，單生花，甚少 2-3 朵簇生。萼片 4-5 片，呈綠色，花瓣 4-5 片，呈白色，雄蕊眾多。

Remarks: Hermaphroditic. Flowers axillary, solitary, seldomly 2-3 clustered. sepals 4-5, green, petals 4-5, white, stamens many.

植物趣聞 ANECDOTE ON PLANTS

外來入侵物種 Invasive Species：

草莓番石榴已經廣泛遍佈全球並成為歸化種，並用於觀賞和農業用途。憑藉快速生長、對環境超群出眾的耐受性和高生產力，草莓番石榴具有極高的競爭力並在環境中迅速成為優勢種。與其他入侵物種一樣，草莓番石榴最初主導着森林的下層林木。然後進一步擴展，最終抵達林冠上層。它們與原生物種競爭陽光和養分，抑制了原生物種的恢復力，導致生物多樣性下降。除了授粉繁殖外，草莓番石榴能透過根莖進行無性繁殖，短時間內建立種群。在這個自然機制下，無形中加劇了其對當地生境的破壞。

草莓番石榴的入侵性強，目前國際自然保護聯盟（IUCN）已將草莓番石榴評級為世界百大外來入侵種之一。從印度洋島嶼到太平洋島嶼和美國，世界各地都有報告其入侵當地的案例。

Strawberry Guava has spread and naturalised across the world for ornamental and agricultural purposes. By virtue of its fast growth, exceptionally unsurpassed tolerance to multiple environments and high productivity, the tree is aggressively competitive propagating across the environments. Comparable with other invasive species, Strawberry Guava initially dominates the understory of forest and continues to expand afterwards. The tree ultimately attains the upper canopy. They compete with the native species for sunlight and nutrient, and inhibit the resilience of native species, thereby resulting in biodiversity loss. Other than pollination, the tree can reproduce asexually through rhizomes. Hence, it can establish a population within a short time. Under this natural mechanism, the damage to local habitats can be intensified.

In respect of its invasiveness, the tree has been rated as one of the 100 World's Worst Alien Species by International Union for Conservation of Nature (IUCN). The invasive cases are reported worldwide, from Indian Ocean islands to Pacific Islands and the USA.

番石榴

Guava | *Psidium guajava* L.

相片拍攝地點：香港動植物公園、川龍
Tree Location: Hong Kong Zoological and
Botanical Gardens, Chuen Lung

名字由來 MEANINGS OF NAME

　　番石榴中的「番」字指該植物由外國引入，而「石榴」則指它的果實形狀與安石榴相似。

　　The common name「番」refers to which the plant is introduced from foreign countries;「石榴」indicates its comparable fruit shape with Pomegranate (*Punica granatum,* 安石榴).

本地分佈狀態 DISTRIBUTIONS	外來物種 Exotic species
原產地 ORIGIN	美洲中部，例如墨西哥。 Central America, as in Mexico.
生長習性 GROWING HABIT	常綠喬木。高度可達 13 米。 Evergreen tree. Up to 13 m tall.

1	2	3	4	5	6	7	8	9	10	11	12	花果期 月份

花期：本港八月至九月。果期：本港八月至九月。
Flowering period: August to September in Hong Kong. Fruiting period: August to September in Hong Kong.

番石榴在市場上廣受歡迎，通常作為水果、果汁和果醬販售。番石榴有兩項天然優勢幫助它遍佈全球各地。首先，它展現出迅速適應不同土壤及氣候的能力。其次，它比其他果樹結果更易更快。從而為業界提供可觀的生產力和收入。與細小苦澀的野生番石榴相比，馴化後的番石榴果實較大且香甜可口，維他命 C 和人體所需礦物質含量尤為豐富。基於以上原因，番石榴樹現遍及熱帶及亞熱帶地區。番石榴的樹形、產量、果實大小和形狀很難保持。為保證銷售的質量，人們傾向使用它地下部分的根莖種植番石榴。透過選擇強壯耐用的根莖，番石榴的良好小枝會嫁接在根莖上，稱之為「接穗」。由於過程中沒有涉及任何有性繁殖，故這方法亦被定義為「營養繁殖」。

Guava is popular in markets and commonly sold as fruit, juice and jam. There are two advantages favouring its dissemination in the world. First, it shows adapts quickly to different environments. Second, it provides faster yields than other fruit trees, thereby offering fascinating productivity and income. Compared with the wild Guava, which the fruits are small and bitter, the domesticated Guava is with fruits that are larger and sweeter, abundant with vitamin C and other essential minerals that required by our body. By these reasons, the tree is now spread throughout the Tropics and the Subtropics. It is hard to maintain a decent tree form, yield, and fruit size and shape for Guava. To preserve the quality for sale, people tend to grow guavas with "rootstocks", which are an underground part of plant. After selecting a strong and durable rootstock, the favourable branchlet of a Guava cultivar is grown on the rootstock, known as "scion". Since mating event is absent, the practice is thus known as "vegetative reproduction".

辨認特徵 TRAITS FOR IDENTIFICATION

① 樹幹 TRUNK	② 樹皮 BARK	③ 葉 LEAVES
④ 花 FLOWERS	⑤ 果 FRUITS	

① 番石榴的樹幹。小枝有棱角，被短柔毛。

Trunk of *Psidium guajava* L. Young branchlets angular, pubescent.

② 樹皮光滑，呈褐黃色，片狀剝落。

Bark smooth, brownish yellow, exfoliating in flakes.

③ 葉對生，呈革質，葉片矩圓形至橢圓形，頂端急尖至鈍形，基部圓形，葉面粗糙，葉背被短柔毛，中脈於葉面下陷，具明顯側脈和網狀紋。

Simple leaves opposite. Blade leathery, oblong to elliptic, apex acute to obtuse, base rounded, adaxially rough, abaxially pubescent, midvein impressed adaxially, lateral and netted veins prominent.

④ 花單生或 2-3 朵組成聚傘花序，腋生。花瓣 4-5 片，白色。雄蕊多數，長度與花柱相若。

Flowers solitary or 2-3 in cymes, axillary. Petals 4-5, white. Stamens many, as long as style.

⑤ 漿果，卵球形或梨形，先端具宿存萼片，成熟時果肉白色及黃色，胎座肥大，肉質，淡紅色。

Berries globose, ovoid, or pyriform, apex with persistent sepals, flesh white or yellow, placenta reddish when mature.

生態 ECOLOGY

　　由於它的果實果汁充沛，能吸引鳥類及其他哺乳類動物，故番石榴常被用作育樹造林及幫助恢復本地生物多樣性之樹種。然而，它在某些地區不但不受歡迎，更被歸類為入侵物種。例如，在加拉巴哥群島，外來番石榴與本地的番石榴屬植物，呈競爭關係。由於這些競爭最終會導致本地物種滅絕，在林地復育引入新物種前，應進行更多的研究。

Since its fruits are juicy and attractive to birds and other mammals, the species always serves as a framework species in reforestation to help restore local biodiverstiy. However, it is not welcomed in some regions and is classified as an invasive species. For example, the introduction of Guava into the Galapagos Islands intensifies the competition with the endemic *Psidium* plant by depriving their resources. The competition could ultimately drive a number of endemic species to extinction. Therefore, more studies should be conducted before introducing a new species for any purposes.

山蒲桃 又稱：白車

Levine's Syzygium | *Syzygium levinei* (Merr.) Merr. & L. M. Perry

相片拍攝地點：沙田公園
Tree Location: Sha Tin Park

名字由來 MEANINGS OF NAME

屬名 *Syzygium* 來自希臘語 *suzugos*，意思是合生。

The generic name *Syzygium* comes from the Greek word *suzugos*, meaning joined.

本地分佈狀態 DISTRIBUTIONS	原生物種 Native species
原產地 ORIGIN	香港和中國東南部，包括廣東、廣西及海南。 Hong Kong and Southeast China, e.g. Guangxi, Guangdong and Hainan.
生長習性 GROWING HABIT	常綠喬木。高度可達 15 米。 Evergreen tree. Up to 15 m tall.

花果期
月份

1	2	3	4	5	6	7	8	9	10	11	12

花期：本港七月至九月。果期：本港二月至五月。
Flowering period: July to September in Hong Kong. Fruiting period: . February to May in Hong Kong.

　　山蒲桃的果實營養豐富，可生食，亦可加工成果泥、果醬和高檔飲料，如葡萄酒。山蒲桃也具高藥用價值，具抗病毒、抗氧化和抗菌作用，常用於外用散瘀消腫、解毒、止血、跌打損傷。

The fruits are rich in nutrients and can be eaten fresh. They can also be processed into fruit paste, jam and high-class beverages like wine. The tree also shows excellent medicinal properties. It is antiviral, antioxidant and antibacterial. The functions of its external application includes relieving blood stasis, bruises and swelling, also promote detoxification and hemostasis.

辨認特徵 TRAITS FOR IDENTIFICATION

① 樹幹 TRUNK	② 樹皮 BARK	③ 葉 LEAVES
④ 花 FLOWERS	⑤ 果 FRUITS	

① 山蒲桃的樹幹。

Trunk of *Syzygium levinei* (Merr.) Merr. & L. M. Perry.

② 樹皮白色。小枝圓柱狀，具糠秕，變乾後轉為灰白色。

Bark whitish. Branchlets terete, chaffy, greyish white when dry.

③ 葉片對生，橢圓形至卵狀橢圓形，革質，基部寬楔形，頂端銳形，葉緣全緣，兩面具光澤，具腺體。側脈與中脈形成 45 度角，具邊脈。

Opposite. Blade elliptic to ovate-elliptic, leathery, base broadly cuneate, apex acute, margin entire, both surfaces glossy, glandular. Secondary veins at an angle of 45° from the midvein, with intramarginal veins.

④ 雌雄同株。圓錐狀聚傘花序頂生或腋生於小枝，花朵眾多，細小，呈白色，雄蕊眾多。

Androgynous. Inflorescences terminal or axillary on apical parts of branchlets, paniculate cymes, numerous flowers, small, white, numerous stamens.

⑤ 果實近球形，7-8 毫米，種子 1 粒。

Fruit subglobose, 7-8mm, 1-seeded.

生態 ECOLOGY

開花時，其花會產生大量花蜜，以吸引動物幫忙傳粉。其果實富含葡萄糖，顏色顯眼，足以吸引以果實為食的動物幫忙傳播種子。

During the flowering seasons, the blossoms yield profuse nectars to attract animals to help pollination. Its fruits are glucose-prolific and with conspicuous colour, which make them attractive to fruit-eating animals that assist seed dispersal.

生命力 VITALITY

山蒲桃通常生長於疏林和溪流附近，而且偏愛日照充足，對乾旱和強風的耐受性較高。

It mostly grows near sparse forests and streams, prefers sunny weather and shows good tolerance to drought and wind.

植物趣聞 ANECDOTE ON PLANTS

蒲桃屬 *Syzygium*：

有時要從外觀大同小異的蒲桃屬中分辨不同種略有難度，就如山蒲桃和蒲桃。在香港，蒲桃被廣泛種植於市區公園和街道，山蒲桃則更常見於近郊和鄉郊。雖然兩者的葉序皆為對生，但蒲桃的葉片更大片，而且花朵和果實的大小都較山蒲桃大。香花蒲桃與山蒲桃的葉片、花和果實大小如出一轍。

Sometimes it is rather difficult to distinguish different species of *Syzygium*, which are similar in appearance, such as Levine's Syzygium and Rose Apple. In Hong Kong, Rose Apple is widely planted in urban parks and streets, while Levine's Syzygium is more commonly found in suburbs and rural areas. Although the phyllotaxy of both is opposite, the leaves, size of flowers and fruits of Rose Apple are larger than that of Levine's Syzygium. The leaves, flowers and fruits of Fragrant Syzygium are exactly the same size as the Levine's Syzygium.

小葉欖仁 又稱：細葉欖仁、非洲欖仁

Madagascar Almond, Terminalia, Umbrella Tree | *Terminalia mantaly* H. Perrier

相片拍攝地點：荔枝角公園、鰂魚涌公園、香港迪士尼樂園
Tree Location: Lai Chi Kok Park, Quarry Bay Park, Hong Kong Disneyland

名字由來 MEANINGS OF NAME

　　Terminalia 在拉丁語中意為葉片在樹枝頂端簇生。此外，小葉欖仁的果實帶有杏仁味，因此別名為「馬達加斯加杏仁」。

The generic name *Terminalia* in Latin describes its leaves crowned at the apex of branches. Its fruits taste like an almond, hence also named as "Madagascar Almond".

本地分佈狀態 DISTRIBUTIONS	外來物種 Exotic species
原產地 ORIGIN	馬達加斯加。 Madagascar.
生長習性 GROWING HABIT	半落葉喬木。高度可達 10 米。 Semi-deciduous tree. Up to 10 m tall.

1	2	3	4	5	6	7	8	9	10	11	12

花果期月份

花期：本港三月至六月。果期：本港四月至九月。
Flowering period: March to June in Hong Kong. Fruiting period: April to September in Hong Kong.

小葉欖仁的葉片、樹皮及木材皆具藥用價值，用以根治痢疾、腹瀉、糖尿病和腸胃炎等疾病。它的木材除可用作燃料外，其木材內含豐富的單寧酸可加工成織物染料。

小葉欖仁擁有筆直而健壯的樹形及傘狀的開放式樹冠，賦予此樹優越的遮蔭效果，故目前此樹種被廣泛種植在公園、花園及街道。

Its leaves, bark and wood are pivotal to treat diseases such as dysentery, diarrhea, diabetes and gastroenteritis. In addition, its wood can be used as fuel and often processed into fabric dyes due to its abundant content of tannin.

Having a handsome tree form and umbrella-shaped open crown, the tree provides excellent shading effect. Therefore, it is widely planted in parks, gardens and street.

辨認特徵 TRAITS FOR IDENTIFICATION

① 樹幹 TRUNK	② 樹皮 BARK	③ 葉 LEAVES
④ 花 FLOWERS	⑤ 果 FRUITS	

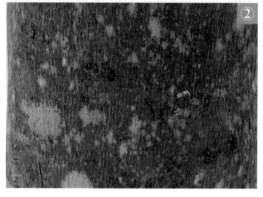

① 小葉欖仁的樹幹。側枝向外水平開展，分層成輪生（假輪生）。

Trunk of *Terminalia mantaly* H. Perrier. Lateral branches explanate, spreading horizontally, tiered into whorls (pseudoverticillate).

② 樹皮黑棕色，筆直，無毛，光滑，具突出褐色皮孔。

Bark blackish brown, erect, glabrous, smooth with prominent brownish lenticels.

③ 葉片常 3-7 片簇生於短枝頂端。葉片紙質，倒卵形，葉緣呈疏鈍鋸齒狀，兩面皆無毛。兩面中脈明顯，側脈不明顯，葉軸兩側具明顯腺體。

Simple leaves often 3-7 crowded at the apex of short branches. Blade papery, glabrous, obovate, margin sparsely crenate, midvein prominent on both surfaces, lateral veins inconspicuous, with apparent glands at axis of lateral veins on both sides.

④ 穗狀花序，花朵細小，呈綠色，花瓣缺失。雄蕊外露，呈乳白色。

Spikes axillary. Flowers small, green, apetalous. Stamens exserted, milky white.

⑤ 核果，光滑，長橢圓形，與橄欖相似，成熟時由綠色轉為暗褐色。

Drupes oblong, smooth, akin to a Chinese olive, turning from green to dark brown when ripe.

植物趣聞 ANECDOTE ON PLANTS

小葉欖仁、大葉欖仁和錦葉欖仁 Madagascar Almond, *Terminalia catappa* (Indian Almond) and *Terminalia mantaly* 'Tricolour'：

小葉欖仁、大葉欖仁和錦葉欖仁皆屬使君子科及欖仁樹屬。要清楚分辨三者，可從葉片大小及葉片顏色入手。把大葉欖仁與小葉欖仁的葉片合併比較，會發現大葉欖仁的葉片比小葉欖仁更為巨大。此外，大葉欖仁的葉片在冬天會轉為紅色，而其他兩種則不會。錦葉欖仁是小葉欖仁的栽培品種，它的葉子為淺綠色，邊緣帶乳白色或乳黃色斑塊，嫩葉則為粉紅色。由於其葉色多變，故被稱為「錦葉欖仁」和「三色小葉欖仁」。

Three species are the members of *Terminalia*, a genus that is derived from the big family Combretaceae. Leaves serves as a key to distinguish the three comparable species. First, the leaves of Indian Almond are rather sizable and those of Madagascar Almond are smaller. Furthermore, the leaves of Indian Almond turn red in winter, whereas the other two do not. Variegated-leaf Almond (*Terminalia mantaly* 'Tricolour') is a cultivar of Madagascar Almond. Its leaves show outstanding pale green with milky white or milky yellow marks on the margin while juvenile leaves are pink. Considering its diverse leaf colours, it is named as "Variegated-leaf Almond" and "Three-colour Small-leaf Almond".

澳洲欖仁

Australian Almond, Mullers Terminalia | *Terminalia muelleri* Benth.

相片拍攝地點：香港迪士尼樂園酒店
Tree Location: Hong Kong Disneyland Hotel

名字由來 MEANINGS OF NAME

　　澳洲欖仁的葉片簇生於枝條頂端，猶如為澳洲欖仁進行加冕儀式，故被冠名為 *Terminalia*，源自拉丁詞 *terminus*（中文意譯為欖仁）。為紀念著名的植物學家 —— 費迪南德‧馮‧穆勒（1825-1896）曾闡明陳述了許多關於植物學學術論文，種加詞特此取為 *muelleri*。

The leaves of the Australian Almond grow in clusters on the top of the branches, like a coronation, so it is named *Terminalia*, which comes from the Latin word *terminus*. The specific epithet *muelleri* is named after Ferdinand von Mueller (1825-1896), a prolific German botanist who contributed numerous botanical academic treatises.

本地分佈狀態 DISTRIBUTIONS	外來物種 Exotic species
原產地 ORIGIN	澳洲北領地和昆士蘭州。 Northern Territory and Queensland, Australia.
生長習性 GROWING HABIT	落葉喬木。高度可達 10 米。 Deciduous tree. Up to 10 m tall.

花果期 月份	1	2	3	4	5	6	7	8	9	10	11	12

花期：本港五月至九月。果期：本港七月至十月。
Flowering period: May to September in Hong Kong. Fruiting period: July to October in Hong Kong.

　　澳洲欖仁葉片茂密，樹冠寬闊，具高觀賞價值。當老葉凋落前，會從綠色轉為紅色，「萬綠叢中一點紅」給這座城市錦上添花。而澳洲欖仁需要單獨種植在開揚的空地上，才能展現出最優美的姿態。

　　Australian Almond is decidedly ornamental due to its dense foliage and extensive horizontal dimension. The leaves turn red when old, embellishing the city with a scenery of "a red spot in the midst of thick foliage". To maintain a decent tree form, Australian Almond is suggested to be planted in solitarily in wide open space.

辨認特徵 TRAITS FOR IDENTIFICATION

① 樹幹 TRUNK	② 樹皮 BARK	③ 葉 LEAVES
④ 花 FLOWERS	⑤ 果 FRUITS	

① 澳洲欖仁的樹幹。莖直立。小枝圓柱狀，堅硬，從樹幹水平處展開。

Trunk of *Terminalia muelleri* Benth. Stem erect. Branchlets terete, stiff, sent out horizontally from the trunk.

② 樹皮微黑暗灰色，無毛。

Bark blackish dark grey, smooth.

③ 互生，聚生於小枝頂端。葉片紙質，鈍形至倒卵形，頂端漸尖，鈍形或圓形，基部楔形，葉緣全緣，葉背葉脈突出。

Simple leaves alternate, crowded at the apex of branchlets. Leaf blade papery, obtuse to obovate, apex acuminate, obtuse or rounded, base cuneate, entire, veins prominent abaxially.

④ 穗狀花序，腋生，雄花與兩性花同株。花朵較小，無柄，花萼 4-5 片，呈白色至粉紅色，沒有花瓣。

Andromonoecious. Spikes axillary. Flowers small, sessile, calyx 4-5, white to pink, apetalous.

⑤ 核果卵形至橢圓形，形似橄欖，呈肉質，成熟時轉為紫色或紫紅色。

Drupes fleshy, ovoid to ellipsoid, akin to a Chinese olive, turning purple or purplish red when ripe.

生態 ECOLOGY

　　澳洲欖仁在乾旱和多風的環境下仍能茁壯成長，澳洲的野生澳洲欖仁分佈於低於海平面 100 米的季風林、海岸森林和沙丘。除古巴外，暫時沒有其他國家遭受澳洲欖仁入侵。

Australian Almond shows a great resilience to drought and windy environments, with its wild population spreading in monsoon forest, coastal forest and sand dunes at elevations below 100 m from sea level in Australia. No country but only Cuba has reported the invasiveness of the tree.

植物趣聞 ANECDOTE ON PLANTS

欖仁樹屬 Relatives of *Terminalia*：

　　欖仁樹屬具眾多高觀賞價值的樹種，包括小葉欖仁、欖仁樹和錦葉欖仁，故被廣泛種植在公園、花園和街道。澳洲欖仁可算是香港最稀有的欖仁樹屬樹種，只會偶爾遇見零星幾棵種植於私營部門，如香港迪士尼樂園。

Terminalia is a genus composed of many ornamental trees, including Madagascar Almond (*T. mantaly*), Indian Almond (*T. catappa*) and *T. mantaly* 'Tricolour'. They are widely planted in parks, gardens and streets. In Hong Kong, Australian Almond is likely the rarest *Terminalia*, occasionally planted by private sectors such as Hong Kong Disneyland.

八角楓

Chinese Alangium, China Alangium | *Alangium chinense* (Lour.) Harms

相片拍攝地點：香港動植物公園、香港中文大學、香港仔樹木研習徑
Tree Location: Hong Kong Zoological and Botanical Gardens, The Chinese University of Hong Kong, Aberdeen Tree Walk

名字由來 MEANINGS OF NAME

　　種加詞 *chinense* 意指中國的，暗示八角楓原生於中國。由於八角楓幼時的葉片有時呈八角形，故得名為「八角楓」。

　　The specific epithet *chinense* means Chinese in Latin, referring to its native distribution in China. Since the blades of its saplings are sometimes octagonal, it is also named as 「八角楓」.

本地分佈狀態 DISTRIBUTIONS	原生物種 Native species
原產地 ORIGIN	八角楓分佈廣泛，從喀麥隆至埃塞俄比亞東南部、南熱帶非洲以及亞熱帶亞洲皆有分佈。 The tree has a broad range of distribution from Cameroon to Southeast Ethiopia and South Tropical Africa, and Subtropical Asia.
生長習性 GROWING HABIT	落葉小喬木或灌木。高度 3 至 5 米，最高可達 15 米。 Deciduous small tree or shrub. 3-5m, up to 15m tall.

1	2	3	4	5	6	7	8	9	10	11	12

花果期月份

花期：本港五月至七月。果期：本港七月至十一月。
Flowering period: May to July in Hong Kong. Fruiting period: July to November in Hong Kong.

八角楓具有多種藥用價值。其根部、葉片和花朵是重要的傳統中藥，用以治療各種痛症。八角楓的根部蘊含喜樹次鹼和八角楓鹼，具舒筋活絡之效。其葉片含有 β-穀固醇和 β-香樹脂醇乙酸乙酯，能紓緩腫痛。值得一提的是，八角楓雖具藥效，但具一定毒性，過分攝取可引致死亡。曾有一名中國男子因服用過多劑量的八角楓根部而死亡，在此呼籲大眾服用時應先向中醫師諮詢專業意見。

八角楓是一種優質木材，常被用於製作家具，如盒子、椅子和衣櫃。此外，其樹皮纖維十分堅韌，常用於製作織繩。

Chinese Alangium is valued for its versatile medicinal functions. Its roots, leaves and flowers are the major sources of traditional Chinese medicines for promoting health. The roots contain venoterpine and dl-anabasine that can activate meridians and soothe sinew. The leaves are an excellent source of β-sitosterol and β-amyrin acetate which can relieve swelling and pain. Notably, despite of its great medicinal effects, it is poisonous and lethal when over-consumed. A Chinese man died from an overdose of the roots of Chinese Alangium, so it is advised to consult a Chinese medicine practitioner before taking it.

Chinese Alangium's wood is a quality timber and is demanded for making boxes, chairs, and wardrobes. Its bark fibres are stiff enough for making ropes.

辨認特徵 TRAITS FOR IDENTIFICATION

① 樹幹 TRUNK	② 樹皮 BARK	③ 葉 LEAVES
④ 花 FLOWERS	⑤ 果 FRUITS	

① 八角楓的樹幹。小枝微呈「之」字形，幼時被短柔毛，漸變無毛。

Trunk of *Alangium chinense* (Lour.) Harms. Branchlets slightly zigzagged, pubescent when young, soon glabrescent.

② 樹皮褐色，平滑，具皮孔。

Bark brown, smooth, lenticellate.

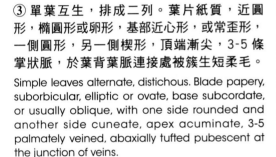

③ 單葉互生，排成二列。葉片紙質，近圓形，橢圓形或卵形，基部近心形，或常歪形，一側圓形，另一側楔形，頂端漸尖，3-5 條掌狀脈，於葉背葉脈連接處被簇生短柔毛。

Simple leaves alternate, distichous. Blade papery, suborbicular, elliptic or ovate, base subcordate, or usually oblique, with one side rounded and another side cuneate, apex acuminate, 3-5 palmately veined, abaxially tufted pubescent at the junction of veins.

④ 聚傘花序，腋生，花朵眾多。花瓣 6-8 片，初時呈白色，其後轉為黃色。雄蕊 6-8 條，輪狀排列，較花瓣長。

Cymes axillary, many-flowered. Petals 6-8, turning from white to yellow later. Stamens 6-8, arranged in whorls, as longer than petals.

⑤ 核果卵球形，成熟時由綠色轉為黑色。

Drupes ovoid, turning from green to black at maturity.

小果鐵冬青 又稱：微果冬青、救必應

Small-fruited Holly, Chinese Holly | *Ilex rotunda* Thunb. var. *microcarpa* (Lindl. ex Paxton) S. Y. Hu

相片拍攝地點：香港中文大學
Tree Location: The Chinese University of Hong Kong

名字由來 MEANINGS OF NAME

　　小果鐵冬青是由開創世界植物學發展的傑出植物學家 —— 胡秀英教授命名。學名中的 S. Y. Hu 是胡教授的簡稱，對她發現該物種表示尊重。俗名「小果」和變種加詞 *microcarpa* 意指其與鐵冬青相比較為細小的果實。

　　Small-fruited Holly is named by Prof. Shiu-Ying Hu, who was an eminent botanist ushering the development of botany in the world. The author's name "S. Y. Hu" in the scientific name is the abbreviation of Prof. Hu, to respect her findings of the species. The common name "Small-fruited" and the variety name *microcarpa* are to describe its smaller fruit size when compared with *I. rotunda*.

本地分佈狀態 DISTRIBUTIONS	原生物種 Native species
原產地 ORIGIN	華南及香港，廣泛分佈於香港島、大帽山及大埔滘。 South China and Hong Kong, with wide distribution in Hong Kong Island, Tai Mo Shan and Tai Po Kau.
生長習性 GROWING HABIT	常綠喬木。高度可達 20 米。 Evergreen tree. Up to 20 m tall.

花果期月份

1	2	3	4	5	6	7	8	9	10	11	12

花期：本港三月至五月。果期：本港十二月至二月。
Flowering period: March to May in Hong Kong. Fruiting period: December to February in Hong Kong.

① 樹幹 TRUNK	② 樹皮 BARK	③ 葉 LEAVES
④ 花 FLOWERS	⑤ 果 FRUITS	

① 小果鐵冬青的樹幹。或具板根。

Trunk of *Ilex rotunda* Thunb. var. *microcarpa* (Lindl. ex Paxton) S.Y. Hu. Sometimes with buttress root.

② 樹皮灰色或灰黑色，光滑。

Bark grey or greyish black, smooth.

③ 單葉互生。葉片革質，無毛，長橢圓狀橢圓形，基部鈍形，甚少圓形或楔形，下延，頂端漸尖，葉緣全緣。中脈於葉背凹陷，於葉背突起。

Simple leaves alternate. Blade leathery, glabrous, oblong-elliptic, base obtuse, rarely rounded or cuneate, decurrent, apex acuminate, margin entire. Midvein impressed adaxially, elevated abaxially.

④ 傘形花序，腋生，3-13 朵花，呈白色，細小，花瓣 5-7 片。總花梗及花梗被微柔毛。雄花具 5-7 條雄蕊。雌雄異株。

Umbels axillary, 3-13-flowered. Flowers white, small, petals 5-7. Peduncles and pedicels puberulous. Male flowers stamens 5-7. Dioecious.

⑤ 核果球形，有時卵形或橢圓形，細小，成熟時轉為紅色。分核 6-7 粒，橢圓形，具三棱，二溝於葉背。

Drupes globose, sometimes ovoid or ellipsoid, small, turning red when mature. Pyrenes 6-7, elliptic, 3-striate, 2-sulcate on the dorsal surface.

植物趣聞 ANECDOTE ON PLANTS

世界著名植物學家 —— 胡秀英
The prominent botanist in the world — Shiu-Ying Hu：

胡教授是中國著名的植物學家，她畢生致力研究冬青屬、泡桐屬、萱草屬、蘭科和菊科。作為一位學富五車、知識豐裕的學者，她發表了 160 多篇學術論文和 185,000 件標本，當中包括了大約 200 種冬青科標本，佔所有冬青科標本的一半。

胡教授於 1968 年受聘於香港中文大學擔任高級講師，並隨即在大學中創辦植物標本室（現名為胡秀英植物標本館），供教學及研究之用。胡教授於 2001 年獲香港政府頒授銅紫荊星章，並於 2002 年獲中大頒授榮譽院士銜，以表揚她對植物科學的傑出貢獻。直至 2012 年，胡教授與世長辭，享年 102 歲。

Prof. Hu was a well-known Chinese botanist, devoting her entire career to studying *Ilex*, *Paulownia*, *Hemerocallis*, *Orchidaceae* and Asteraceae. As a prolific scholar, she published more than 160 academic treatises and 185,000 specimens, including specimens of around 200 species of Aquifoliaceae, one half of the specimens of Aquifoliaceae.

Prof. Hu accepted the position as a senior lecturer in the Chinese University of Hong Kong (CUHK) in 1968 and soon initiated a herbarium (currently named as Shiu-Ying Hu Herbarium) in the university for teaching and research purposes. Prof. Hu was awarded by the Hong Kong Government with the Bronze Bauhinia Star in 2001, and was successively bestowed the honorary fellowship by CUHK in 2002 for her notable contributions to plant science. In 2012, Prof. Hu passed away at the age of 102.

冬青屬的多樣性 Variety of *Ilex*：

雖然《中國植物誌》和全球生物多樣性資訊機構（GBIF）偏向承認鐵冬青多於小果鐵冬青，但我們仍可以指出「兩種」之間的外形差異（是否「兩種」目前仍有爭議）。要區分兩者，可從果實大小、總花梗和花梗入手。小果鐵冬青的果實較細小，總花梗和花梗被微柔毛，鐵冬青的果實較大，總花梗和花梗無毛。

Although *Flora of China* and Global Biodiversity Information Facility (GBIF) prefer to accept *I. rotunda* more than *I. rotunda* var. *microcarpa*, we can tell the morphological differences between "two species" (quotation for the controversial status). The most distinct characteristics between them are which Small-fruited Holly has smaller fruits and puberulous peduncles and pedicels, whereas the *I. rotunda* has larger fruits and glabrous peduncles and pedicels.

重陽木

Chinese Bishopwood | *Bischofia polycarpa* (H. Lév.) Airy Shaw

相片拍攝地點：港鐵火炭站對出
Tree Location: Next to Fo Tan MTR Station

名字由來 MEANINGS OF NAME

不同神話演譯了重陽木的起源。中文名稱「重陽木」正象徵中國的傳統節日 —— 重陽節。傳聞過去人們在重陽節祭祖登高、緬懷先人期間，發現重陽木樹冠闊大，可以提供最佳的遮蔭效果。因此被命名為「重陽木」。

另一個重陽木名字由來的説法是因其生命週期長。「九九重陽節」（重陽節的別稱）中「九」與「久」同音，同時九是中國最大的個位數，象徵長壽。因此它被賦予了與節日相對應的名稱「重陽木」。

The Chinese common name of Chinese Bishopwood is「重陽木」, showing greatly pertinent to the Chung Yeung Festival, a traditional festival in China. Rumour has it that people in the past realized that a Chinese Bishopwood was able to confer unsurpassed shading effect for hikers during the Chung Yeung Festival, hence giving it a credit to「重陽木」.

本地分佈狀態 DISTRIBUTIONS	外來物種 Exotic species
原產地 ORIGIN	中國華南秦嶺、淮河流域，至華北福建、廣東等省份。 North-central, South-central and Southeast China.
生長習性 GROWING HABIT	落葉喬木。高度可達 15 米。 Deciduous tree. Up to 15 m tall.

1	2	3	4	5	6	7	8	9	10	11	12	花果期月份

花期：本港四月至五月。果期：本港十月至十一月。
Flowering period: April to May in Hong Kong. Fruiting period: October to November in Hong Kong.

Another rumour about the naming of the tree is related to the longevity of Chinese Bishopwood. The tree is appreciated for its rather longer lifespan which fortuitously matches the meaning of Double-Ninth Festival (a nickname of the Chung Yeung Festival). Nine is the largest single digit number and is emblematic of longevity in Chinese culture. Therefore, the tree is named as「重陽木」for being pertinent to the festival.

應用 APPLICATION

重陽木的木材細膩、堅韌且抗真菌,常用於製作家具和橋樑。其葉片具藥用價值,有消腫、解毒、活血之效。它也是蒸餾酒和葡萄酒的常用成分。此外,從種子提煉的油分可用於生產潤滑劑或肥皂。

The wood of Chinese Bishopwood is fine, tough and antifungal; thus, is demanded for making furniture and bridges. The leaves are applied as traditional Chinese medicines to disperse swelling, detoxification and promote blood circulation; they are also a prevalent ingredient of distilled liquors and wines. Furthermore, oil extracted from the seeds can be processed into lubricant and soap.

辨認特徵 TRAITS FOR IDENTIFICATION

① 樹幹 TRUNK	② 樹皮 BARK	③ 葉 LEAVES
④ 花 FLOWERS	⑤ 果 FRUITS	

① 重陽木的樹幹。

Trunk of *Bischofia polycarpa* (H. Lév.) Airy Shaw.

② 樹皮褐色,縱裂。

Bark brown colour, longitudinally cracked.

③ 掌狀三出複葉,葉片紙質,卵形或橢圓狀卵形,有時長橢圓狀卵形。葉片基部圓鈍形或淺心形,葉緣具鈍細鋸齒,頂端銳形或短漸尖,頂端葉片常大於兩側葉片。

Palmately ternate. Blade papery, ovate or elliptic-ovate, rarely oblong-ovate, base rounded-obtuse to shallowly cordate, margin obtusely serrulate, apex acute to shortly acuminate, terminal leaflet relatively larger.

④ 花單性，雌雄異株。花序着生於新枝下方，纖細，下垂。雄花花瓣 5 片。雌花萼片與雄花相同，但具白色膜質邊緣。

Dioecious. Inflorescences growing on new branches, slender, drooping. Male flowers sepals 5. Male flowers pistilloid eminent. The sepals of the female flowers are the same as those of the male flowers, sepals with white membranous margin.

⑤ 漿果，無毛，成熟時轉為棕紅色。

Berries globose, chestnut at maturity.

生態 ECOLOGY

重陽木的果實可食用，味道酸甜可口，雀鳥皆好之。

The fruit of Chinese Bishopwood is edible, sweet and sour, birds are falling for it.

植物趣聞 ANECDOTE ON PLANTS

重陽木與秋楓 Chinese Bishopwood and *Bischofia javanica* (Autumn Maple)：

重陽木與秋楓是香港市區常見的行道樹及觀賞樹，但如果缺少深入觀察，便會很容易將兩者混淆。要清楚分辨兩者，可從樹皮、葉片及花序入手。

重陽木的樹皮呈褐色帶縱裂，而秋楓的樹皮則呈灰棕色，沒有裂紋。此外，重陽木的葉緣為鈍鋸齒狀，基部為圓鈍形或淺心形，而秋楓則為淺鋸齒狀，基部為闊楔形。重陽木的花序為總狀花序，而秋楓則為圓錐花序。

Chinese Bishopwood and Autumn Maple are widely cultivated in Hong Kong as street and ornamental trees. They are outwardly indistinguishable but can be unmistakably identified from barks, leaves and flowers.

First, the bark of Chinese Bishopwood is brown and cracked longitudinally, whereas the one of Autumn Maple is greyish brown without longitudinal crack, but rather peeling off in flakes. Second, the blade of Chinese Bishopwood contains obtusely serrate margin and rounded-obtuse to shallowly cordate base, whereas that of Autumn Maple shows shallowly serrate margin and broadly cuneate base. Third, the inflorescences of Chinese Bishopwood are racemes, while those of Autumn Maple are panicles.

山烏桕

Mountain Tallow Tree | *Sapium discolor* (Champ. ex Benth.) Müll. Arg.

相片拍攝地點：香港中文大學
Tree Location: The Chinese Unversity of Hong Kong

應用 APPLICATION

山烏桕的根部和葉片具藥用價值，其根部能利尿通便，消腫散結，解蛇蟲毒。它的葉片可以活血化瘀，解毒利濕。此外，種子上的蠟質層可用於製作肥皂和蠟燭。

The roots and leaves are valued for their wonderful medicinal properties; the roots can promote diuresis, relieve swelling and stasis, and treat snake bug poisoning; the leaves are effective to encourage blood circulation and detoxification, and attenuate dampness. In addition, the wax coating on seeds is exploited for making soaps and candles.

本地分佈狀態 DISTRIBUTIONS	原生物種 Native species
原產地 ORIGIN	中國南部和其他東南亞國家，如印度、泰國、老撾、尼泊爾、菲律賓和越南。 Southern China and the countries of Southeast Asia, eg. India, Thailand, Laos, Nepal, Philippines and Vietnam.
生長習性 GROWING HABIT	半落葉灌木至小喬木。高度可達 12 米。 Semi-deciduous shrub or small tree. Up to 12 m tall.

花果期月份	1	2	3	4	5	6	7	8	9	10	11	12

花期：本港四月至六月。果期：本港七月至十月。
Flowering period: April to June in Hong Kong. Fruiting period: July to October in Hong Kong.

① 樹幹 TRUNK	② 樹皮 BARK	③ 葉 LEAVES
④ 花 FLOWERS	⑤ 果 FRUITS	

① 山烏桕的樹幹。

Trunk of *Sapium discolor* (Champ. ex Benth.) Müll. Arg.

② 樹皮光滑，灰棕色，具皮孔。

Bark smooth, grey-brown, lenticellate.

③ 單葉互生，葉片紙質，橢圓形或長橢圓狀卵形，頂端鈍形或短漸尖，基部楔形，成熟時由紅色轉為綠色，冬天落葉前轉為紅色。葉柄頂端具 2 毗連的腺體。

Simple leaves alternate. Blade papery, elliptic or oblong-ovate, apex obtuse or shortly acuminate, base cuneate, young leaves and aging leaves red. Petioles with 2 glands at apex.

④ 雌雄同株，單性花。總狀花序頂生，呈黃色，雄花生於花序軸上半，雌花則生於下半。

Monoecious, unisexual flower. Racemes terminal, yellow, with several female flowers in the lower part, many male flowers in the upper part.

⑤ 蒴果球狀，蠟質，成熟時轉為黑色。
Capsules globose, waxy, black when ripe.

生態 ECOLOGY

　　山烏桕是蝴蝶和飛蛾幼蟲的寄主植物。例如，癩皮瘤蛾的幼蟲以山烏桕的葉片為食。此外，山烏桕葉柄頂端具產糖腺體，同時亦被稱為花外蜜腺，這是大戟科的常見特徵，用以吸引螞蟻，利用牠們趕走草食性動物。此外，山烏桕的果實美味多汁，對紅耳鵯、白頭鵯和松鼠等本地動物具高吸引力。

Mountain Tallow Tree is the host plants for butterfly and moth larvae. For example, its leaves are the primary food source of *Gadirtha inexacta's* larvae. Besides, there is a pair of glands near the leaf base secreting nectars. It is a common feature of Euphorbiaceae. This feature can attract ants, which are hired as "bodyguards" to drive away any herbivore. In addition, its fruits are juicy and serve as a delicacy to many local animals such as Red-whiskered Bulbul, Light-vented Bulbul and squirrels.

生命力 VITALITY

　　山烏桕廣泛分佈於香港的次生林和灌叢中。它是一種早期植林的先鋒物種，可以在濃蔭或其他不同生長環境中茁壯成長。雖然這棵樹能適應不同棲息地，但由於其在貧瘠和全日照的裸露地區生長表現未如理想，故它並非荒地的理想先鋒樹種。

By virtue of its excellent resilience to multiple environments, the tree is a pioneer species for rehabilitation in Hong Kong. Whilst Mountain Tallow Tree is rather undemanding to different habitats, it shows a poor performance in extremely barren and overexploited lands. Therefore, it is not an ideal pioneer species.

木油樹 又稱：千年桐、皺桐
Wood-oil Tree | *Vernicia montana* Lour.

相片拍攝地點：川龍、大埔滘、城門標本林
Tree Location: Chuen Lung, Tai Po Kau, Shing Mun Arboretum

名字由來 MEANINGS OF NAME

　　種加詞 *montana* 意謂其原生地為山上；由於該樹可長年產油，中文名稱「千年桐」是指它具有長遠的經濟價值；而其眾所周知具皺摺的樹皮，也使它擁有另一中文名稱叫「皺桐」。

The specific epithet *montana* describes its native range on mountains. The common name "Wood-oil Tree" refers to its oil production; the tree can yield oils for years, hence named as「千年桐」for highlighting its rather prolonged economic value. As all know enclosed in a wrinkled bark, it is also offered a Chinese name "Wrinkled Tung".

本地分佈狀態 DISTRIBUTIONS	外來物種 Exotic species
原產地 ORIGIN	中國東至西南部省份，也分佈於越南、泰國和緬甸等地。 The provinces of East and Southwest China, also distributed in Vietnam, Thailand and Myanmar.
生長習性 GROWING HABIT	常綠或落葉喬木。高度可達 20 米。 Evergreen or deciduous tree. Up to 20 m tall.

1	2	3	4	5	6	7	8	9	10	11	12	花果期 月份

花期：本港四月至六月。果期：本港七月至十月。
Flowering period: April to June in Hong Kong. Fruiting period: July to October in Hong Kong.

　　油桐屬種子的油分經處理後可提煉成桐油。一般而言，市場上的桐油提煉自木油樹或油桐（桐樹）。由於提煉出來的油為乾性油，具輕盈、不導電及防水的特性，它曾被用作船舶塗漆，但現已被油漆取替。木油樹的葉片和根部被用作傳統中藥，以根治潰瘍、類風濕性關節炎、促進消化和減少體內寄生蟲。此物種的木材呈白色、具光澤、無特殊氣味而且經久耐用，因此常用於製作家具。鑑於其樹冠闊大，大樹可乘涼，故目前此樹種在街道上被廣泛種植。

The seeds of *Vernicia* species are oily, and the extracts are the primary source of Tung oil. Tung oil in markets generally refers to the extracts from Wood-oil Tree or Tung Tree (*Vernicia fordii*). Since extracted oil is drying oil, which is light, nonconductive and waterproof, it was used for ship painting but has soon been substituted by paints which are more economical. Wood-oil Tree's leaves and roots are effective traditional Chinese medicines to treat ulcers and rheumatoid arthritis, promote digestion, and reduce internal parasites from the body. The timbers are white, glossy, without special smell, and long-lasting; therefore, it is yearned for making furniture and other constructions. In addition, the species is a widely planted street tree by virtue of its large crown and shading effect.

辨認特徵 TRAITS FOR IDENTIFICATION

① 樹幹 TRUNK	② 樹皮 BARK	③ 葉 LEAVES
④ 花 FLOWERS	⑤ 果 FRUITS	

① 木油樹的樹幹。

Trunk of *Vernicia montana* Lour.

② 樹皮呈褐色，小枝無毛，具稀疏隆起的皮孔。

Bark brown, branchlets yellowish pubescent, glabrescent, lenticellate.

③ 互生，葉片寬卵形，基部心形至截形，頂端急尖至漸尖，葉緣淺裂 2-5 片或全緣，葉脈呈掌狀。葉竇或葉柄頂端鄰常具杯狀線體。

Simple leaves alternate. Broadly ovate, base cordate to truncate, apex acute to acuminate, margin 2-5 lobed or entire, palmately veined. A pair of sizable cupular glands at the apex of petiole and sinus.

④ 單性花，雌雄異株，稀少雌雄同株。雄花：花瓣 5 片，底部呈白和紫紅色，具紫紅色條紋，雄蕊 8 至 10 條。雌花：花萼和花瓣與雄性花形態相同，花柱 3 條，二深裂，呈 Y 狀，黃綠色。

Unisexual flower, dioecious, rarely monoecious. Male flowers: petals 5, white and sometimes crimson at base, with purplish red stripes, stamens 8-10. Female flowers: calyx and petals as male flowers, styles 3, bipartite, Y-shaped, yellow-green.

⑤ 核果，卵圓形，體積大，具三棱。果皮革質和具網狀皺紋。

Drupes ovoid, large volume, 3-ribbed. Pericarp leathery, with reticulate wrinkles.

生態 ECOLOGY

其結實和舒展伸延的根部能抓緊泥土，防止水土流失。

Its strong and stretching roots could hold soil and prevent soil erosion.

生命力 VITALITY

木油樹能抵抗病蟲害，並偏好生長於排水良好及透氣的土壤。

The tree is relatively tolerant to diseases and pests, and prefers soils with good drainage and aeration.

荔枝

Lychee, Lichee | *Litchi chinensis* Sonn.

相片拍攝地點：元朗公園、老圍村
Tree Location: Yuen Long Park, Lo Wai Village

原產地 ORIGIN

荔枝分佈廣泛，大部分野生荔枝分佈在不同中國省份，例如雲南和四川。現時，我們普遍見到的荔枝都已經被馴化。近年，科學家揭開了荔枝歷史神祕的面紗，他們認為早熟品種來自雲南，晚熟品種則來自海南。栽培品種的擴散間接促成雜交，意味着混合最少兩種相似品種會使新的雜交種誕生。「妃子笑」是一種在由雲南品種和海南品種培育出的受歡迎的雜交品種。

Lychee is one of the most distributed fruit trees in the world, with its origin from China, e.g. Yunnan and Sichuan. The tree we see nowadays is actually a product under a long journey of domestication. Recently, scientists have discovered the origins of two progenitors of Lychee, with the early-maturing cultivar from Yunnan and the late-maturing cultivar from Hainan. The spread of cultivars indirectly promotes hybridization, meaning the mixing of at least two similar species will lead to the birth of new hybrid. "Smiling Concubines" is the popular hybrid of Yunnan's and Hainan's cultivars.

本地分佈狀態 DISTRIBUTIONS	外來物種 Exotic species
生長習性 GROWING HABIT	常綠喬木。高度可達 10 米。 Evergreen tree. Up to 10 m tall.

花果期 月份

1	2	3	4	5	6	7	8	9	10	11	12

花期：本港二月至四月。果期：本港五月至六月。
Flowering period: February to April in Hong Kong. Fruiting period: May to June in Hong Kong.

名字由來 MEANINGS OF NAME

荔枝因連枝採摘，故名為「離枝」。從東漢開始，該物種才被改稱為同音的「荔枝」。

Lychee was previously named as「離枝」due to which the fruits are always harvested with branches attached. Until the Eastern Han (25-220 AD), the tree has been renamed into「荔枝」, which is the homophone of「離枝」in Chinese.

應用 APPLICATION

荔枝可加工成荔枝乾及荔枝酒。其木材亦適合製作家具。

Lychee can be processed into dried lychee and lychee wine. Its wood is suitable for making furniture.

辨認特徵 TRAITS FOR IDENTIFICATION

① 樹幹 TRUNK	② 樹皮 BARK	③ 葉 LEAVES
④ 花 FLOWERS	⑤ 果 FRUITS	

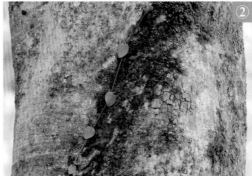

① 荔枝的樹幹。

Trunk of *Litchi chinensis* Sonn.

② 樹皮稍粗糙，灰黑色，初時棕紅色。

Bark scabrous, grayish black, initially reddish brown.

③ 偶數羽狀複葉，小葉 2-4 對近對生。葉片被針形或卵狀被針形，有時長橢圓狀被針形。基部楔形至圓形，頂端驟尖或短尾狀漸尖，薄革質或革質。葉背粉綠色，被白色粉末，中脈明顯突起，側脈間有距離。

Paripinnately compound alternate, leaflets 2-4 pairs, subopposite. Leaf blade lanceolate or ovate-lanceolate, sometimes long elliptic-lanceolate. Base cuneate to rounded, apex shortly caudate-acuminate or cuspidate, thinly leathery or leathery. Back pink-green, with white powder, midrib prominent, lateral veins apart.

④ 圓錐花序頂生，多分枝，大，花萼金色被
絨毛。雄花雄蕊 6-8。雌花子房密被糙硬毛
和具瘤。

Panicles terminal, multi-branched, large, calyx
golden tomentose. Male flowers stamens 6-8.
Ovaries of female flowers densely hispid and
tuberculous.

⑤ 果皮成熟時為暗紅至鮮紅色，球狀到近球
形，具小瘤。種子被擬假種皮包裹。

Fruits globose to subglobose, pericarp usually
dark red to fresh red at maturity, tuberculous.
Seeds covered by arillode.

植物趣聞 ANECDOTE ON PLANTS

荔枝的歷史 History of Lychee：

　　荔枝是中國著名的果樹品種，幾千年
來古今帝王們為之着迷。荔枝的歷史可
追溯至漢朝。漢武帝征服廣東後，將荔
枝等亞熱帶、熱帶水果引進上林苑的扶
荔宮，作為世界上熱帶植物園的雛型。
可惜，果樹無法適應當地氣候而死亡。
但是漢武帝並無因此而放棄，反而萌生
利用快馬從中國南方運來荔枝，以保持
荔枝的新鮮度的想法。荔枝容易變質。
白居易在《荔枝圖序》曾曰：「若離本
枝，一日而色變，二日而香變，三日而
味變，四五日外，色香味盡去矣」，意為
保存荔枝之路困難重重。在唐代，楊貴
妃鐘愛荔枝，唐玄宗為博美人一笑，命
下人快馬加鞭將新鮮荔枝送達皇宮。

　　Lychee is a prominent fruit tree in China and
fascinated emperors for millennia. The earliest
record of Lychee was documented in the Han
dynasty. Followed by Emperor Wu of Han (141 BC-
87 BC) conquering Guangdong, he introduced
Lychee and other subtropical and tropical fruits into
Fuli Palace in Shanglinyuan, which was probably
the most preliminary tropical botanical garden in
the world. However, the introduced Lychee failed to
acclimatise to the environment. Emperor Wu of Han
did not give up and ignited an idea of delivering
the fresh fruits from Southern China to his palace by
fast horse for a sake of the preservation of freshness.
The fruits spoil rapidly; as Bai Juyi mentioned in his
preface of《荔枝圖序》:「若離本枝，一日而色變，
二日而香變，三日而味變，四五日外，色香味盡去
矣」, indicating the difficulty of preserving the fruits.
In the Tang Dynasty, Yang Guifei was an aficionado
of Lychee. Wanting to please her, Emperor
Xuanzong of Tang Dynasty ordered his servants to
deliver fresh lychees to the palace quickly.

無患子 又稱：木患子、肥皂果

Soap Berry | *Sapindus saponaria* L.

相片拍攝地點：東涌北公園、香港中文大學
Tree Location: Tung Chung North Park, The Chinese Unversity of Hong Kong

名字由來 MEANINGS OF NAME

屬名 *Sapindus* 源自拉丁詞 *Sapo* 及 *indus*，分別帶有肥皂和印度的含意，泛指來自印度含豐富皂苷的果實。除了以 *saponaria*（中文意譯為肥皂）作其種加詞，亦有 Soap Berry 的俗名（中文意譯為肥皂果），意指每顆果實都猶如一塊「小肥皂」。

The generic name *Sapindus* is a blend of the Latin words *Sapo* (soap) and *indus* (India or River Indus), collectively meaning Indian soap. The specific epithet *saponaria* and the common name "Soap Berry" refer to its fruits are like small soaps..

本地分佈狀態 DISTRIBUTIONS	原生物種 Native species
原產地 ORIGIN	原生於熱帶和亞熱帶美洲，亦廣泛分佈於中國、日本、韓國及印度。 Native to tropical and subtropical America. The tree is also widely distributed in China, Japan, Korea and India.
生長習性 GROWING HABIT	落葉喬木。高度可達 20 米。 Deciduous tree. Up to 20 m tall.

1	2	3	4	5	6	7	8	9	10	11	12

花果期月份

花期：本港三月至五月。果期：本港六月至十二月。
Flowering period: March to May in Hong Kong. Fruiting period: June to December in Hong Kong.

應用 APPLICATION

無患子因其藥用價值而備受重視。無患子的果皮經曬乾加工後，便是常用中藥「無患子皮」。無患子皮具有清熱化痰止咳之效。無患子的根部則可加工成「無患樹蔃」，內服後能治外感發熱，並排出積存體內的熱毒。

由於無患子的果皮蘊含豐富皂苷，故亦被稱為肥皂果。果皮泡水搓揉後會產生泡沫，用作肥皂。古時，人們常常收集果皮，用以製作各種天然無害的洗滌用品，如洗滌劑、洗髮水和肥皂。此外，根據研究顯示，皂苷具顯著抗真菌性，而且具有控制真菌傳播的潛力，對未來醫學和環境層面都能作出貢獻。

由於無患子樹形雄偉、線條優美，故常被用作觀賞樹種植。

Soap Berry has been exploited for multiple medicinal merits. The dried fruit peels are the primary source of a traditional Chinese medicine, *Wu Huan Zi Pi*, which shows wonderful effects on relieving heat, phlegm and cough. The roots are processed into *Wu Huan Shu Qiang* that can expel persisted internal and external heat.

The name of Soap Berry is derived from its saponin-rich fruit peels. Its fruit peels can act as soap after soaking and rubbing. In the past, people collected the fruit peels to make different kinds of natural and harmless cleaning products, such as detergents, shampoos and soaps. Besides, according to some studies, saponin has remarkable antifungal properties and the potential to control the spread of fungi, which can contribute to the medical and environmental aspects in the future.

Soap Berry is commonly planted as an ornamental tree due to its beautiful tree form.

辨認特徵 TRAITS FOR IDENTIFICATION

① 樹幹 TRUNK	② 樹皮 BARK	③ 葉 LEAVES
④ 花 FLOWERS	⑤ 果 FRUITS	

① 無患子的樹幹。

Trunk of *Sapindus saponaria* L.

② 樹皮灰棕色或黑棕色，光滑，片狀剝落。幼枝呈綠色，無毛，具皮孔。

Bark greyish brown or blackish brown, smooth, peeling off in flakes. Young branches green, glabrous, lenticellate.

③ 偶數羽狀複葉，互生，小葉 5-8 對，對生或近對生。葉軸腹面具凹槽，無毛或被短柔毛。小葉薄紙質，狹橢圓狀披針形或稍鐮形，基部楔形，稍偏斜，葉緣全緣，兩面皆無毛，側脈密集，細長。

Paripinnately compound alternate, leaflets 5-8 pairs, opposite or subopposite. Rachis grooved adaxially, glabrous or pilosulose. Blade thinly papery, narrowly elliptic-lanceolate or slightly falcate, base cuneate, slightly oblique, margin entire, hairless on both sides, lateral veins dense, slender.

④ 圓錐花序頂生，雌雄同株。花朵較小，被短柔毛，花瓣 5 片，輻射對稱，呈乳白色。

Monecious. Panicles terminal. Flowers small, pubescent, petal 5, actinomorphic, milky.

⑤ 可育分果呈橙色，近球形，1 粒種子，乾時轉為黑色並具皺摺，不育分果 1-2 粒，著生在可育分果頂端。

Fertile schizocarps orange, subglobose, 1-seed, black and wrinkled when dry. Infertile schizocarps 1-2, bearing at the apex of the fertile schizocarp.

生命力 VITALITY

無患子對乾旱土壤、炎熱及寒冷的環境具良好耐受性。

It shows good tolerance to arid soils, and hot and cold environments.

槭樹科

嶺南槭

Tutcher's Maple | *Acer tutcheri* Duthie

相片拍攝地點：城門標木林
Tree Location: Shing Mun Arboretum

名字由來 MEANINGS OF NAME

香港前林務監督德邱（1868-1920）於 1894 年在大嶼山首次發現嶺南槭，並用他的名字命名。嶺南槭的模式標本是從香港採集。因為嶺南槭數量稀少，故被列入香港珍稀植物（第 4 類）。

W. J. Tutcher (1868-1920) was credited with his first observation of Tutcher's Maple from Lantau Island in 1894. Its type specimen is collected in Hong Kong. It is listed in Rare and Precious Plants of Hong Kong (category 4) for valuing its small number.

應用 APPLICATION

嶺南槭秋天時葉色艷麗，果實則色澤紅潤，它在中國被廣泛種植在花園和公園中。

The leaf colour of Tutcher's Maple is decorative in the autumn and its fruit is red. As a result, it is always planted in China as a visual alternative to evergreen trees in gardens and parks.

本地分佈狀態 DISTRIBUTIONS	原生物種 Native species
原產地 ORIGIN	原生於香港、廣東、廣西、福建、江西、浙江及湖南。 Hong Kong, Guangdong, Guangxi, Fujian, Jiangxi, Zhejiang and Hunan.
生長習性 GROWING HABIT	落葉喬木。高度可達 10 米。 Deciduous tree. Up to 10 m tall.

花果期月份	1	2	3	4	5	6	7	8	9	10	11	12

花期：本港四月。果期：本港九月。
Flowering period: April in Hong Kong. Fruiting period: September in Hong Kong.

| ① 樹幹 TRUNK | ② 樹皮 BARK | ③ 葉 LEAVES |
| ④ 果 FRUITS | | |

① 嶺南槭的樹幹。

Trunk of *Acer tutcheri* Duthie.

② 樹皮光滑，呈褐色或暗褐色。小枝初時綠色，無毛。

Bark brown or dark brown, smooth. Branchlets purplish green when immature, glabrous.

③ 葉片對生。葉片膜質，具 3-5 裂，葉緣具細鋸齒，無毛，有時葉背脈腋被毛，3 條主脈。

Simple leaves opposite. blade membranous, broadly ovate, usually 3-5 lobed, lobes margin sparsely serrulate, base rounded to subtruncate, apex acuminate, glabrous, sometimes tufts at the junctions of abaxial veins, basal veins 3.

④ 翅果，成熟時由紅色轉為黃色，其翅幾乎完全水平展開，就像一個「人」字。

Samara turning from red to yellowish at maturity, wings spreading acutely or nearly horizontally, arranged in a Chinese word「人」.

註：本樹另有雜性花。圓錐花序，頂生。花瓣 4 片，呈淡黃白色，雄蕊 8 條，呈紅色。

Remarks: Andromonoecious. Panicles terminal. Petals 4, pale yellowish white, stamens 8, red.

生態 ECOLOGY

　　嶺南槭分佈在疏林中，在茂密的樹冠中甚少存活。嶺南槭的果實顏色吸引眾多昆蟲。同時，其翅果獨特的形狀，使它在缺乏動物傳播的情況下仍能傳播開去。

Tutcher's Maple is distributed in thin forests where trees do not form in crowded canopy. The fruit colour of it attracts a lot of insects. The fruit is winged and allows it to catch a breeze and travel for a long distance without any aid of animals.

植物趣聞 ANECDOTE ON PLANTS

嶺南槭和楓香 Tutcher's Maple and *Liquidambar formosana* (Chinese Sweet Gum)：

　　嶺南槭和楓香形態相似，兩者皆具 3 裂葉形，如缺少深入觀察，很容易將兩者混淆。要清楚分辨兩者，可從葉序及果實入手。楓香的葉序為互生，其果實為球狀蒴果；而嶺南槭的葉序為對生，果實是翅果。值得留意的是，嶺南槭只在香港某些地區常見，如馬鞍山和大東山。因此，您在市區看到的大多數具 3 裂葉形的樹木可能都是楓香。

Tutcher's Maple is outwardly imperceptible to Chinese Sweet Gum with similar leaf shape in 3 lobes. Without careful observation, it is difficult to distinguish the two. To distinguish the trees, the leaves of Chinese Sweet Gum are arranged alternately, whereas those of Tutcher's Maple are opposite. On the contrary of the fruits of Chinese Sweet Gum which are globose capsules with needle like appendages, those of Tutcher's Maple are samaras. It is worth noting that the wild Tutcher's Maple is biogeographically confined to certain areas such as Ma On Shan and Sunset Peak in Hong Kong. Therefore, most of the trees with 3-lobed leaves that we see in the urban areas are more likely Chinese Sweet Gum.

人面子 又稱：銀蓮果

Yanmin | *Dracontomelon duperreanum* Pierre

相片拍攝地點：城門谷公園、 元朗公園、迪欣湖活動中心
Tree Location: Shing Mun Valley Park, Yuen Long Park, Inspiration Lake Recreation Centre

名字由來 MEANINGS OF NAME

屬名 *Dracontomelon* 源自拉丁詞 *drakon* 及 *melon*，用以形容瓜狀的果實。俗名 Yanmin 中文音譯為「人面」，意指人面子的果實內果皮的細孔酷似人臉。

The generic name *Dracontomelon* is a blend of the Latin words *drakon* and *melon*, collectively delineating its melon-like fruits. The common name "Yanmin" can be directly translated into「人面」in Chinese, referring to its endocarp with pores akin to a human face.

應用 APPLICATION

人面子的果實、根部和葉片皆具有藥用價值。其果實是傳統中藥中的「人面子」，具健脾解毒之效。其根皮搜集作「人面子根皮」，可用於治療皮膚疾病。「人面子葉」則治療癤瘡。

本地分佈狀態 DISTRIBUTIONS	外來物種 Exotic species
原產地 ORIGIN	中國中南部和東南部、緬甸和越南。 South-Central and Southeast China, Myanmar and Vietnam.
生長習性 GROWING HABIT	常綠喬木。高度可達 25 米。 Evergreen Tree. Up to 25 m tall.

1	2	3	4	5	6	7	8	9	10	11	12

花果期月份

花期：本港四月至五月。果期：本港六月至十一月。
Flowering period: April to May in Hong Kong. Fruiting period: June to November in Hong Kong.

除藥用價值外，其果實還可食用，常被加工成醃製水果和中式醋栗醬。其木材緊密、具光澤，而且耐用、具顯著防腐能力，故被廣泛採用作家具製作和建材。其種子油可加工成肥皂和潤滑油。

人面子擁有壯觀的圓形樹冠，常被視為觀賞樹種廣泛種植在公園和花園。

Yanmin has been exploited for multiple medicinal functions with its fruits, roots and leaves. The fruits are the traditional Chinese medicine *Ren mian zi*; it can effectively invigorate spleen and promote detoxification. The root barks are collected for *Ren mian zi gen pi* for treating skin diseases. The leaves are processed into *Ren mian zi ye* for treating furunculosis.

Apart from medicinal versatility, the fruits are edible and often processed into pickled fruits and Chinese gooseberry sauce for seasoning. The heartwood is dense, shiny and durable, with notable rot-resistant ability; therefore, it is highly yearned for making furniture and constructions. The extracted seed oil can be processed into soap and lubricant.

By virtue of its handsome tree crown in rounded shape, the tree is often planted as a shade in parks and gardens.

辨認特徵 TRAITS FOR IDENTIFICATION

① 樹幹 TRUNK	② 樹皮 BARK	③ 葉 LEAVES
④ 花 FLOWERS	⑤ 果 FRUITS	

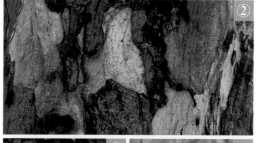

① 人面子的樹幹。具板根。小枝具條紋，被細小灰色絨毛。

Trunk of *Dracontomelon duperreanum* Pierre. Buttressed root. Branchlets striped, minutely grey tomentose.

② 樹皮光滑，呈灰白色或黃綠色，片狀剝落。

Bark smooth, greyish white or greenish yellow, peeling off in flakes.

③ 奇數羽狀複葉，互生，小葉 11-17 片，互生。小葉片近革質，長橢圓形至長橢圓狀披針形，葉背具明顯網狀脈，脈腋處被白色簇絨。

Imparipinnately compound alternate, leaflets 11-17 alternate. Blade subleathery, oblong to oblong lanceolate, reticulate venation prominent and white tufts at axils of lateral veins abaxially.

④ 雌雄同體。圓錐花序頂生，花朵呈黃綠色，被微絨毛。

Hermaphroditic. Panicles terminal. Flowers yellowish green, tomentulose, petals white.

⑤ 核果成熟時由綠色轉為黃色，內果皮扁平，先端具 4-5 個小孔。

Drupes green to yellow when mature, endocarp compressed, with 4-5 pores at apex.

人面子對乾旱、潮濕、強風、蟲害和污染物具中等耐受性。

Yanmin is moderately tolerant to drought, dampness, strong wind, pests and pollutants.

古樹名木 Old and Valuable Trees（OVT）：

有 3 棵人面子被列入香港古樹名木，分別位於元朗公園和梅樹坑。目前最大的人面子位於梅樹坑（編號：LANDSD TP/2），胸徑約 1274 毫米，高度約 28 米，樹冠約 27 米。

In Hong Kong, 3 trees of Yanmin are registered as OVTs and they are located in Yuen Long Park and Mui Shue Hang, respectively. Yanmin in Mui Shue Hang (LANDSD TP/2) is currently considered as the largest, with a DBH of 1274 mm, a height of 28 m and a crown spread of 27 m.

人面子古樹名木

嶺南酸棗
Ling Nan Suan Zao | *Spondias lakonensis* Pierre

相片拍攝地點：屯門市廣場對出
Tree Location: Next to Tuen Mun Town Plaza

本地分佈狀態 DISTRIBUTIONS	外來物種 Exotic species
原產地 ORIGIN	中國中南部、東南部，同時亦分佈於泰國、老撾、越南和馬來亞。 South-Central and Southeast China, Thailand, Laos, Vietnam and Malaya.
生長習性 GROWING HABIT	落葉喬木。高度可達 15 米。 Deciduous tree. Up to 15 m tall.

花果期 月份	1	2	3	4	5	6	7	8	9	10	11	12

花期：本港四月至六月。果期：本港八月至十月。
Flowering period: April to June in Hong Kong. Fruiting period: August to October in Hong Kong.

① 樹幹 TRUNK	② 樹皮 BARK	③ 葉 LEAVES
④ 花 FLOWERS	⑤ 果 FRUITS	

① 嶺南酸棗的樹幹。成熟時具板根。

Trunk of *Spondias lakonensis* Pierre. Roots buttressed when mature.

② 樹皮呈灰棕色。

Bark greyish brown.

③ 奇數羽狀複葉，互生，小葉 11-23 對，對生至互生，葉柄及葉軸稍被短柔毛。小葉片長橢圓形或長橢圓狀披針形，頂端漸尖，基部歪形、寬楔形至圓形，葉緣全緣，沒有邊脈，葉背沿葉脈微被短柔毛。

Imparipinnately compound alternate, leaflets 11-23 pairs, opposite to alternate. Petioles and rachis minutely pubescent. Blade oblong or oblong-lanceolate, apex acuminate, base oblique, broadly cuneate to rounded, margin entire, without intramarginal veins, minutely pubescent abaxially along veins.

④ 圓錐花序，腋生，大型。花瓣呈白色，開花時反曲。雌雄同體。

Panicles axillary, large. Petals white, recurved when flowering. Hermaphroditic.

⑤ 核果卵形至倒卵形，具角，成熟時轉為紅色。

Drupes ovate to obovate, angled, turning red when mature.

嶺南酸棗的分類學簡史 A brief taxonomic history of *Spondias*：

　　檳榔青屬曾是漆樹科下的一個大屬。然而，鑑於種間形態差異，此屬的代表性受到不少植物分類學家所質疑。故隨後檳榔青屬被分成兩群，分別為原來的檳榔青屬與新增的南酸棗屬。與此同時，部分原檳榔青屬的物種亦被安排至相對應的屬。例如，南酸棗的學名 *Spondias axillaris* 變更為 *Choerospondias axillaris*。嶺南酸棗因缺乏邊脈且有一些形態上的差異，故歸類至檳榔青屬。

　　分類學是科學家們孜孜不倦、鍥而不捨地逐步摸索未知領域的心血結晶。憑藉一眾科學家鉅細無遺地將各種生物按其特徵結構進行分類，從古老的形態學至現代分子生物學，一一從中整理出有用的資訊，將進化的奧祕以先進前衛的科技呈現於眾人眼前。

Spondias used to be a large genus in Anacardiaceae. Since the genus was in dispute of its validity to represent all the members it covered, the genus has been further fragmented into two groups, namely the original genus *Spondias* and the newly added genus *Choerospondias*. At the same time, some species of the original *Spondias* were also arranged to the corresponding genus. For example, Hog Plum, previously named as *Spondias axillaris*, has been rectified into *Choerospondias axillaris*. Ling Nan Suan Zao has been rearranged to *Spondias*, due to the absence of intramarginal veins and other subtle morphological differences.

Taxonomy is the brainchild of many scientists and a lot of efforts have been involved in investigating the unknown areas. The scientists have categorized comprehensively the organisms according to their characteristics and sorted out useful information from ancient morphology to modern molecular biology. They depicts the mystery of evolution by using the state-of-art technology.

香港的嶺南酸棗 Ling Nan Suan Zao in Hong Kong：

　　嶺南酸棗的樹冠緊湊，白花仿如漫天的繁星，樹上的果實成為它青綠色裙子上的裝飾，倍感秀麗。然而，嶺南酸棗在香港較為罕見，實際上，現有的嶺南酸棗大多數是在引入火焰木及人面子時混雜其中，這兩種樹是香港普遍的行道樹或觀賞樹。

Ling Nan Suan Zao is an excellent ornamental tree with its compact crown, showy white blossoms and decorative fruits. However, it is relatively rare in Hong Kong. In fact, most of the existing Ling Nan Suan Zao are mixed with *Spathodea campanulata* (African Tulip Tree) and *Dracontomelon duperreanum* (Yanmin), which are common street trees or ornamental trees in Hong Kong.

常綠臭椿 又稱：福氏臭椿
Ailanthus, Green Ailanthus | *Ailanthus fordii* Noot.

相片拍攝地點：荔枝角公園、美孚新邨
Tree Location: Lai Chi Kok Park, Mei Foo Sun Chuen

名字由來 MEANINGS OF NAME

前政府花園部監督的查爾斯·福特（1844-1927）於 1884 年至 1886 年期間在香港鶴咀半島採集了常綠臭椿的第一份模式標本，種加詞 *fordii* 和俗名「福氏臭椿」中的「福氏」是為了表彰他發現該物種的貢獻。由於雄花總是散發出臭氣熏天的氣味，因此該品種也被稱為「常綠臭椿」，「臭」是指它散發的氣味。

The specific epithet *fordii* and the common name "Ford's Ailanthus" are in honour of Charles Ford (1844-1927), a superintendent of the Government Gardens from 1884 to 1886, who collected the type specimen of Ailanthus in Cape D'Aguilar in Hong Kong. In respect of the pungent male flowers, Ailanthus in Chinese is also named as「常綠臭椿」while「臭」refers to odour.

本地分佈狀態 DISTRIBUTIONS	原生物種 Native species
原產地 ORIGIN	中國南方和西南的省份，如廣東和雲南。 The provinces of South and Southwestern China, including Guangdong and Yunnan.
生長習性 GROWING HABIT	常綠喬木。高度可達 15 米。 Evergreen tree. Up to 15 m tall.

1	2	3	4	5	6	7	8	9	10	11	12

花果期月份

花期：本港十月至十一月。果期：本港十二月至四月。
Flowering period: October to November in Hong Kong. Fruiting period: December to April in Hong Kong.

① 樹幹 TRUNK	② 樹皮 BARK	③ 葉 LEAVES
④ 花 FLOWERS	⑤ 果 FRUITS	

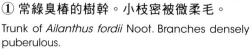

① 常綠臭椿的樹幹。小枝密被微柔毛。

Trunk of *Ailanthus fordii* Noot. Branches densely puberulous.

② 樹皮呈灰棕色，具橫向條紋。

Bark greyish brown with transverse stripes.

③ 偶數羽狀複葉，簇生於枝條頂端，稍傘狀，小葉 6-13 對，對生或近對生。葉片紙質至革質，長橢圓狀卵形，基部鈍形或歪形，頂端短漸尖或鈍形，葉緣波狀，葉背散生細小而平坦的腺點，葉柄具脊。

Paripinnately compound, clustered at the apex of branches and slightly umbrella-shaped, leaflets 6-13 pairs, opposite or subopposite. Blade papery to leathery, oblong-ovate, base obtuse or oblique, apex short acuminate or obtuse, undulate, small and flat scattered glands abaxially. Petiolule ridged.

④ 單性花或雜性花，雌雄同株。圓錐花序頂生，呈淡黃至白色。

Unisexual or polygamous, monoecious. Panicle terminal, large, pale yellow to white.

⑤ 翅果，懸垂葉下，呈狹長葉狀。成熟時由淡綠色轉為褐色。種子藏在翅果中心。

Samaras drooping beneath the leaves, narrow-long and leaf-like. Turning from pale green to brownish at maturity. Seeds are hidden in the center of samaras.

應用 APPLICATION

　　常綠臭椿很高，有令人驚嘆的樹冠，能營造優越的遮蔭效果，故在街道上廣泛種植。常綠臭椿的樹形結構容許風在樹冠底下穿透，因而改善行人路的通風環境。此外，常綠臭椿具高觀賞及教育價值，是公園常見樹種。

In respect of its towering height and dense foliage, the tree is widely planted on streets for providing unsurpassed shady effect. In addition, its tree structure allows penetration of wind in under-canopy areas and hence ameliorates ventilation at pedestrian level. Also, the evergreen Ailanthus has high ornamental and educational value and is a common tree species in the parks.

植物趣聞 ANECDOTE ON PLANTS

價值 Value：

　　常綠臭椿在《香港稀有及珍貴植物》中被評為「近危」。根據《林務規例》第96章，在香港出售、要約售賣或管有常綠臭椿的任何部分即屬違法。為提高公眾對自然保育的意識，常綠臭椿在香港被積極廣泛培植。成功的保育工作令常綠臭椿現時在香港無處不在。

This tree is rated as Near Threatened in *Rare and Precious Plants of Hong Kong* and highly protected due to its ubiquity in Hong Kong. According to the legislation of the Forests and Countryside Ordinance (Cap. 96), it is illegal to sell, offer for sale, or possess any of its portions in Hong Kong. In order to raise public awareness for promoting nature conservation, the tree is broadly cultivated in Hong Kong as an ornamental and street tree species.

桃花心木 又稱：小葉桃花心木

West Indies Mahogany, Cuba Mahogany | *Swietenia mahagoni* (L.) Jacq.

相片拍攝地點：香港墳場
Tree Location: Hong Kong Cemetery

名字由來 MEANINGS OF NAME

屬名 *Swietenia* 以 Gerard von Swieten（1700-1772）的名字命名，他是奧地利女大公瑪麗婭·特蕾西婭的私人醫生。17 世紀時，來自西非的約魯巴人被賣至牙買加作奴隸。他們把另一棵同樣原生於非洲的非洲楝與桃花心木搞混。他們稱這棵樹為 *M'Oganwo* 和 *M'Ogani*，後來美國人把它翻譯成「桃花心木」。這就是桃花心木拉丁學名 *mahagoni* 的由來。

桃花心木主要以葉片大小來與大葉桃花心木作區別。由於桃花心木擁有較為細小的葉片，它亦被稱為「小葉桃花心木」。

本地分佈狀態 DISTRIBUTIONS	外來物種 Exotic species
原產地 ORIGIN	原生於美洲南部至中部。它目前廣泛種植在中國華南和馬來西亞等熱帶地區。 Native to Southern and Central America. It is now widely cultivated in tropical areas such as Southern China and Malaysia.
生長習性 GROWING HABIT	常綠或落葉喬木。高度可達 25 米。 Evergreen or deciduous tree. Up to 25 m tall.

花果期 月份	1	2	3	4	5	6	7	8	9	10	11	12

花期：本港五月至六月。果期：本港十月至十一月。
Flowering period: May to June in Hong Kong. Fruiting period: October to November in Hong Kong.

The generic name *Swietenia* is named after Gerard von Swieten (1700-1772), who was the personal physician of Maria Theresa, Archduchess of Austria. The specific epithet *mahagoni* is derived from the Yoruba vernacular. During the 17th century, many Yoruba people were traded from Nigeria to Jamaica as slaves. They confused the native West Indies Mahogany with (*Khaya senegalensis*) in their home country, and intuitively referred the trees to M'Oganwo and *M'Ogani*. The American later translated the tree name into "Mahogany".

The tree is distinguished from Honduran Mahogany (*Swietenia macrophylla*) through the leaf size. It is named as 「小葉桃花心木」as the leaf size of it is smaller. 「小葉」means small leaves in Chinese.

應用 APPLICATION

　　桃花心木的心材呈紅棕色，有堅固、具光澤、抗蟲害、耐用且易於加工的優點。因此，它被廣泛應用在船隻、家具及藝術品製造，是炙手可熱的木材品種。桃花心木是很多富麗堂皇的歐洲建築之原材料。如位於多明尼加共和國的歐洲殖民建築聖瑪麗亞‧拉梅諾爾大教堂中的木雕就是以此木材製成，在 500 年後仍歷久彌新！

The heartwood is dyed in elegant chestnut. It is sturdy, lustrous, durable and ease-of-processing; therefore, it is always considered as an extravagant timber for making ships, furniture and artworks. The timber also serves as the primary material of many majestic European buildings. For instance, the carvings in the Cathedral of Santa María la Menor in the Dominican Republic are made of West Indies Mahogany and have been already preserved for 500 years!

辨認特徵 TRAITS FOR IDENTIFICATION

① 樹幹 TRUNK	② 樹皮 BARK	③ 葉 LEAVES
④ 花 FLOWERS		

① 桃花心木的樹幹。有板根。

Trunk of *Swietenia mahagoni* (L.) Jacq. Roots buttressed.

② 樹幹結實，樹皮呈微紅色，鱗片狀。枝條光滑，呈灰色。

Trunk sturdy, bark slightly red, scale-like. Branchlets grey, smooth.

③ 羽狀複葉，互生，小葉 8-12 片，對生，螺旋狀排列於小枝上。小葉葉片卵形至披針形，基部歪形，葉緣全緣，葉面呈暗綠色，葉背則呈淡綠色。

Paripinnately compound alternate, leaflets 8-12 opposite, arranged spirally on the branches. Blade ovate to lanceolate, base very oblique, margin entire, adaxially dark green, abaxially light green.

④ 雌雄同株，單性花，圓錐花序腋生。花朵細小，花瓣 5 片，呈綠白色。

Monecious, unisexual flowers. Panicles axillary. Flowers small, petals 5, greenish white.

註：本樹另有蒴果，呈褐色，卵形，木質。種子具翅。

Remarks: Capsule brown, ovoid, woody. Seeds winged.

生態 ECOLOGY

　　由於非法砍伐和過度開採，桃花心木目前被列為瀕危物種。為保育此物種，樹木的採伐和貿易活動受到《瀕危野生動植物種國際貿易公約》（CITES）附錄二的監管和約束。

The wild West Indies Mahogany is currently threatened from rampant logging events and habitat destruction. To conserve its wild population, the logging and trading of the tree is now regulated by the appendix II of the Convention on International Trade in Endangered Species of Wild Fauna and Flora (CITES).

香椿

Chinese Mahogany, Chinese Toona | *Toona sinensis* (A. Juss.) M. Roem.

相片拍攝地點：祐民街、香港動植物公園
Tree Location: Yau Man Street, Hong Kong Zoological and Botanical Gardens

名字由來 MEANINGS OF NAME

香椿樹高壯觀、枝葉繁多且長壽，故名「椿」。因為香椿長壽，故人們常用「椿」字向長輩祝壽。「千椿」和「椿壽」皆為祝福別人萬壽無疆。「椿」亦是中國人的姓氏。傳聞孔子的兒子孔鯉為了不打擾父親思考，故經過院子時會小步快速地安靜走過，即「趨庭而過」。當中「庭」字取自古時尊稱父親「椿庭」。古人也用一種百合科植物 —— 萱草比喻母親，「椿萱並茂」用以形容父母健在。種加詞 *sinensis* 描述了它起源於中國。

本地分佈狀態 DISTRIBUTIONS	外來物種 Exotic species
原產地 ORIGIN	原生於中國中北部、中南部和東南部省份，同時亦分佈於喜馬拉雅山東部和東南亞國家，如越南、老撾和泰國。 Native to provinces of North-Central, South-Central and Southeast China, and the tree is also distributed in the Eastern Himalaya and Southeast Asian countries, such as Vietnam, Laos and Thailand.
生長習性 GROWING HABIT	落葉喬木。高度可達 25 米。 Deciduous tree. Up to 25 m tall.

1	2	3	4	5	6	7	8	9	10	11	12	花果期 月份

花期：本港六月至八月。果期：本港十月至十二月。
Flowering period: June to August in Hong Kong. Fruiting period: October to December in Hong Kong.

Chinese Mahogany can grow to a lofty height and survive for ages. Therefore, it has been always bonded to longevity in Chinese culture. For example,「千椿」and「椿壽」are words for blessing someone to have a longer life.「椿」is also a family name. Rumour has it that Confucius's son tiptoed quietly across the yard (庭) away from his father's room, to just not disturb him from thinking. As a result, there is a saying「趨庭而過」, and the word「庭」is taken form「椿庭」, the term used to call father in the ancient time to show respect. Since *Hemerocallis fulva* (Fulvous Day-Lily, 萱草), which is a species in Liliaceae, metaphorically referred to mother in the past, the Chinese used the term「椿 (father) 萱 (mother) 並 (both) 茂 (thrive)」to describe that their parents are still healthy and alive. The specific epithet *sinensis* describes its origin from China.

應用 APPLICATION

香椿全株皆具藥用價值。其幼葉是一道令人垂涎欲滴的佳餚，同時亦是一種傳統中藥材「椿葉」，具治療噁心、嘔吐、食慾不振和腹瀉等之效。香椿的花可用以治療久咳；果實「香椿子」則可紓緩外感風寒及痢疾。香椿木材強韌，具顯著防腐能力，故被廣泛採用作家具製作。

Chinese Mahogany serves a host of medicinal functions. The young leaves are a delicacy and the primary source of *Chunye*, a traditional Chinese medicine which is effective to treat nausea, vomiting, poor appetite and diarrhea. The flowers can attenuate long-lasting coughing and the fruits *Xiangchunzi* can mitigate exogenous wind-cold and dysentery. Its wood is an outstanding material for making furniture by virtue of its great flexibility, strength and resistance to decay.

辨認特徵 TRAITS FOR IDENTIFICATION

① 樹幹 TRUNK	② 樹皮 BARK	③ 葉 LEAVES
④ 花 FLOWERS	⑤ 果 FRUITS	

① 香椿的樹幹，具板根，受損時釋出大蒜和胡椒氣味。

Trunk of *Toona sinensis* (A. Juss.) M. Roem. Roots buttressed. Garlic and pepper smells when damaged.

② 樹皮灰色至暗褐色，粗糙，片狀剝落。

Bark grey to dark brown, coarse, peeling off in flakes.

③ 偶數羽狀複葉，對生或互生，小葉 14-28 片，對生。小葉葉片紙質，卵狀披針形至長圓狀橢圓形，基部不對稱，頂端尾形至銳形，葉緣具疏鋸齒或甚少全緣，葉背中脈及側脈凸起，側脈 18-24 對。

Paripinnately compound, opposite or alternate, leaflets 14-28, opposite. Blade papery, glabrous, ovate-lanceolate to oblong-elliptic, base asymmetric, apex caudate to acute, margin sparsely serrate or rarely entire, midveins and lateral veins raised abaxially, and lateral veins 18-24 pairs.

④ 雌雄花在同一花序。圓錐花序與葉片長度相約，或較葉片長，腋生。花朵較小，花瓣 5 片，呈白色或粉紅色。雌雄同株。

Female and male flowers on the same inflorescence. Panicles as long as leaves or even longer, axillary. Flowers small, calyx cupular, 5-lobed or undulate, petals 5, white or pink. Monoecious.

⑤ 蒴果橢圓形或卵形，成熟時裂成 5 裂片，暗褐色。種子呈褐色，上端具膜質長翅。

Capsules narrowly elliptic or ovoid, dark brown, splitting into 5 valves when ripe. Seeds brown, with long membranous wing at the upper side.

植物趣聞 ANECDOTE ON PLANTS

古樹名木 Old and Valuable Tree (OVT)：

位於香港動植物公園的香椿（編號：LCSD CW/53）因擁有約 870 毫米顯著的胸徑及約 17 米的高度，故被列入香港古樹名木。

The Chinese Mahogany in Hong Kong Zoological and Botanical Gardens has been registered as a OVT (LCSD CW/53) due to its significant diameter at breast height (DBH) of 870 mm, and a height of 17 m.

香椿古樹名木

柚子

Pomelo, Pummelo | *Citrus maxima* (Burm.) Merr.

相片拍攝地點：獅子會自然教育中心
Tree Location: Lions Nature Education Centre

名字由來 MEANINGS OF NAME

種加詞 *maxima* 的意指為「最大」，代表其果實與其他柑橘屬植物相比較大。

The specific epithet *maxima* means "the biggest", referring to its rather sizable fruit when compared with other *Citrus* spp.

本地分佈狀態 DISTRIBUTIONS	外來物種 Exotic species
原產地 ORIGIN	緬甸北部。 Northern Myanmar.
生長習性 GROWING HABIT	常綠喬木。高度可達 10 米。 Evergreen tree. Up to 10 m tall.

花果期 月份	1	2	3	4	5	6	7	8	9	10	11	12

花期：本港四月至五月。果期：本港九月至十二月。
Flowering period: April to May in Hong Kong. Fruiting period: September to December in Hong Kong.

作為全世界皆有分佈的果樹，柚子因其果實富含維他命 C 及 B、類胡蘿蔔素、黃酮類化合物及其他副產品而受到重視。除了可生食外，亦可加工成沙律、甜品及飲料。

其葉片密被油點，碾碎後會散發柑橘味芬香。俗語有云：「年廿九，洗碌柚」，華人習慣於農曆年廿九用柚子葉洗澡，以「洗走」霉運。

Being one of the global available fruit trees, Pomelo's fruit is valued for its copious nutrient content which is constituted mainly by vitamins C and B, carotenoids, flavonoids and other secondary products. Other than eating it fresh, the fruits can be also processed into salads, desserts and beverages.

Its leaves are densely covered with oil glands and release citrus fragrance when crushed. The Chinese people get used to taking a bath with pomelo leaves when Lunar New Year is approaching to ward off evil from last year.

| ① 樹幹 TRUNK | ② 樹皮 BARK | ③ 葉 LEAVES |
| ④ 花 FLOWERS | ⑤ 果 FRUITS | |

① 柚子的樹幹。小枝扁平，具棱，密被短柔毛，具刺或無刺。

Trunk of Citrus maxima (Burm.) Merr. Branchlets flat and ridged, densely pubescent, spiny or not spiny.

② 樹皮褐色，光滑。

Bark brown, smooth.

③ 單身複葉，互生。葉柄具翅。小葉葉片厚革質，初時密被短柔毛，葉面呈暗綠色，兩面具黃色腺點，卵形至橢圓形，基部圓形，頂端鈍形或圓形，葉緣全緣。

Unifoliate compound alternate. Petioles winged. Blade thick leathery, densely pubescent when young, adaxially dark green, yellow glands on both surfaces, ovate to elliptic, base rounded, apex obtuse or rounded, entire.

④ 雌雄同體。總狀花序，腋生，稀單生，芳香。花瓣 4-5 片，呈淡紫紅色或乳白色，雄蕊眾多，花柱粗長。

Hermaphroditic. Racemes axillary, or rarely solitary, fragrant. Petals 4-5, pale purplish red or milky white, stamens many, style thick and long.

⑤ 柑果球形至闊圓錐形，果皮厚、海綿質，成熟時由綠色轉為淡黃色。

Hesperidia globose to broadly conical, pericarp thick and spongy, turning green to pale yellow when mature.

植物趣聞 ANECDOTE ON PLANTS

柑橘屬的簡介 A brief introduction of *Citrus*：

柑橘屬由 156 個物種組成。經過多重雜交和物競天擇，該屬的親緣關係狀態遠較預想中複雜，多年來困擾眾多科學家。生物技術經歷重重障礙後成功發展，並於基因組學的角度為柑橘屬帶來新的突破。近日一項基因組學研究顯示，目前的柑橘屬物種可能源自 *C. linczangensis* ——一種原始柑橘屬物種，其化石標本曾於喜馬拉雅山脈的東南山麓被發現。*C. linczangensis* 被懷疑曾於中新世晚期經歷了快速種化，並演化為橘子、柚子和枸櫞，分別分佈於印度東北部、緬甸北部和雲南西北部。柑橘屬的三個祖先透過自然和人工雜交成檸檬（由酸橙和枸櫞雜交而成）、甜橙（由柚子和橘子雜交而成）、葡萄柚（由柚子和甜橙雜交而成）。有趣的事，柚子的基因被認為是柑橘屬中最為優勢，顯示其參與性高，常被選擇成雜交品種。與眾多雜交種相較，葡萄柚與柚子的基因相似度最高，約有 63% 的基因源自柚子。

Citrus is a large genus of 156 species. Due to the drastic hybridization and selection events, the phylogenetic status of the genus is far more complicated than expected and has confused scientists for ages. Under the intractable development of biotechnology, it has opened avenues to disentangle the mystery of *Citrus* from a genomic perspective. A recent genomic study suggests that the current *Citrus* species could be derived from *C. linczangensis*, a primitive *Citrus* species whose fossil specimen was found in southeast foothills of the Himalayas. *C. linczangensis* is suspected to experience rapid speciation events during the late Miocene, and evolved into Mandarin (*C. reticulata*), Pomelo (*C. maxima*) and Citron (*C. medica*) which were distributed respectively in northeastern India, northern Myanmar and northwestern Yunnan. The three progenitors of *Citrus* were further hybridized spontaneously or anthropogenically into Lemon (a hybrid of Sour orange and Citron), Sweet Orange (a hybrid of Pomelo and Mandarin) and Grapefruit (a hybrid of Pomelo and Sweet Orange). Intriguingly, the genes of pomelo are suggested to be most predominant among the genus, indicating its assiduous involvement in hybridization. Among the hybrids, Grapefruit shares the highest genetic similarity with Pomelo, with 63% of the genes are derived from Pomelo.

黃皮 又稱：黃皮子、黃皮果

Wampi, Wampee | *Clausena lansium* (Lour.) Skeels

相片拍攝地點：香港動植物公園、獅子會自然教育中心、賽馬會德華公園、川龍
Tree Location: Hong Kong Zoological and Botanical Gardens, Lions Nature Education Centre, Jockey Club Tak Wah Park, Chuen Lung

名字由來 MEANINGS OF NAME

由於其黃澄澄的外皮及晶瑩剔透的果肉，故被稱為「黃皮」。

Because of its golden skin and crystal clear flesh, it is called「黃皮」(yellow skin) in Chinese.

本地分佈狀態 DISTRIBUTIONS	外來物種 Exotic species
原產地 ORIGIN	原生黃皮遍佈於中國南部省份，如雲南及福建。黃皮主要廣泛種植於熱帶和亞熱帶地區，尤其在南中國。 Provinces of South China, e.g. Yunnan and Fujian. Wampi is widely cultivated in tropical and subtropical regions mainly, especially in Southern China.
生長習性 GROWING HABIT	常綠小喬木。高度可達 12 米。 Evergreen small tree. Up to 12 m tall.

1	2	3	4	5	6	7	8	9	10	11	12	花果期月份

花期：本港四月至五月。果期：本港七月至八月。
Flowering period: April to May in Hong Kong. Fruiting period: July to August in Hong Kong.

237

　　黃皮有各式各樣的烹調方法，包括剝皮生吃、榨汁、加工成果醬和天然生曬黃皮乾。黃皮亦可入藥，「黃皮葉」在中藥的四氣五味中，有苦、辛、平的性質，有解表散熱、順氣化痰之效。「黃皮果」有化痰消食之效。此外，黃皮果實含有香豆素，具有清除 DPPH 自由基和超氧陰離子自由基，有抗氧化作用。

Apart from peeling off the skins and eating the fruits fresh, they are also commonly processed into juice, jams and dried Wampee. The tree has various medical functions. The leaves, known as *Huangpiye*, are a competent traditional Chinese medicine that is with bitter, pungent and neutral nature, and can be used for relieving heat and phlegm. The fruits *Huangpiguo* is effective in resolving food stagnancy and relieving phlegm. The fruits also contain profuse coumarins that have anti-oxidation effect through scavenging DPPH radicals and superoxide anions.

辨認特徵 TRAITS FOR IDENTIFICATION

① 樹幹 TRUNK	② 樹皮 BARK	③ 葉 LEAVES
④ 花 FLOWERS	⑤ 果 FRUITS	

① 黃皮的樹幹。

Trunk of *Clausena lansium* (Lour.) Skeels.

② 樹皮光滑呈暗綠或灰綠色，具縱裂紋。小枝長硬毛，具細小腺點。

Bark smooth, with longitudinal crack, dark green or grey green. Branchlets hirsute, with fine glandular dots.

③ 葉片為奇數羽狀複葉，小葉 5-11 對，互生，小葉卵形或卵狀橢圓形，不對稱，葉緣波狀或具淺圓齒，中脈密被絨毛，葉面常無毛。腺點廣泛散佈於葉面，壓碎時會釋出咖喱味。

Imparipinnately compound alternate, leaflets 5-11 alternative. Blade ovate to ovate-elliptic, asymmetrical, margin undulate or coarsely crenate, leaf rachis densely pubescent, adaxially often glabrous. Scattered glandular dots on both surfaces, curry smelling when crushed.

④ 大型圓錐花序。花瓣 5 片，呈白色，被柔毛。雄蕊 10 條，長度不一。

Large panicles terminal. Petals 5, white, pubescent. Stamens 10, which varying in length.

⑤ 漿果，橢球形或寬卵球形，初時具分散油點。淡黃至暗黃色。

Berry, ellipsoid or broadly ovoid, with scattered oil spots at first, light to dark yellow.

生命力 VITALITY

　　黃皮對不同類型的土壤皆具有良好的耐受性，不過其易受漬澇和乾旱影響，故此偏愛排水良好和肥沃的土壤。

Wampi moderately acclimatises to multiple types of soil, but is vulnerable to waterlogging and drought. Therefore, it prefers well-drained and nutritious soils.

九里香 又稱：千里香、月橘

Orange-jessamine │ *Murraya paniculata* (L.) Jack

相片拍攝地點：沙田公園、柴灣戲院對出
Tree Location: Sha Tin Park, next to Chai Wan Cinema

名字由來 MEANINGS OF NAME

　　種加詞 *paniculata* 指其圓錐花序。中文名稱「九里香」指其花香遠益清。值得一提的是，這棵樹在大陸俗稱「千里香」，在台灣俗稱「月橘」。

本地分佈狀態 DISTRIBUTIONS	外來物種 Exotic species
原產地 ORIGIN	九里香分佈在台灣和南中國省份，例如海南和廣東。 Taiwan and provinces of South China, e.g. Hainan and Guangdong.
生長習性 GROWING HABIT	常綠灌木或小喬木。高度可達 12 米。 Evergreen shrub or small evergreen tree. Up to 12 m tall.

花果期 月份	1	2	3	4	5	6	7	8	9	10	11	12

花期：本港四月至八月。果期：本港九月至十二月。
Flowering period: April to August in Hong Kong. Fruiting period: September to December in Hong Kong.

The specific epithet *paniculata* refers to its paniculate inflorescences. The Chinese name 「九里香」 exaggerates the delightful fragrance of its blossoms that can still smell from miles far. Notably, the tree is commonly named as 「千里香」 in China's mainland and 「月橘」 in Taiwan.

應用 APPLICATION

　　九里香的每部分均可食用且具藥用價值。九里香的葉片對治療胃痛及持續性腹痛有特效，是製作中國民間必備胃藥「39 胃泰」的重要材料。此外，從九里香的葉片和果實中提取的精油具抗炎和鎮痛功用。

　　九里香可在烹飪中使用。花朵可加工成花茶；葉片常用於馬來西亞美食。除了醫藥和食品方面的應用外，九里香對修剪具高耐受性，因此被廣泛用作樹籬和剪型植物。

The whole plant is edible and valued for versatile medicinal functions. First, the leaves are effective to treat gastralgia and stagnant abdominal pain; it is one of the primary ingredients of 「39 胃泰」, a pervasive stomach medicine in China. Moreover, essential oils extracted from the leaves and fruits show anti-inflammatory and analgesic effects.

It can be used in cooking. The flowers can be processed into flowering tea. The leaves are commonly used in Malaysian cuisines. By virtue of its excellent resilience to pruning, the tree is used as hedges and topiary plants.

辨認特徵 TRAITS FOR IDENTIFICATION

① 樹幹 TRUNK	② 樹皮 BARK	③ 葉 LEAVES
④ 花 FLOWERS	⑤ 果 FRUITS	

① 九里香的樹幹。

Trunk of *Murraya paniculata* (L.) Jack.

② 樹皮灰棕色，老時由平滑具光澤變得粗糙。一年生枝條呈綠色。

Bark greyish brown, turning from glossy, smooth to rough when old, annual branches green.

③ 葉片互生，3-5 片，甚少 7 片互生小葉。葉片卵狀披針形至闊倒卵形。基部對稱或歪形，頂端漸尖至圓形，具光澤。

Leaf blade alternate, leaflets 3-5 (-7) alternate. Leaf blade ovate-lanceolate to obovate-elliptic, base symmetrical or oblique, apex long acuminate to obtuse, glossy.

④ 圓錐花序或聚傘花序，頂生或腋生。花瓣 5 片，倒被針形或狹長橢圓形，呈白色，完全盛開時稍為反折。

Panicles or cymes, terminal or axillary. Petals 5, oblanceolate or narrowly elliptic, white, slightly reflexed when full blooming.

⑤ 果實狹橢圓形或甚少卵形。成熟時由橙色轉為朱紅色，果皮上具腺點。

Fruits narrowly ellipsoid, or rarely ovoid, turning from orange to vermilion when mature, pericarp with glandular dots.

生態 ECOLOGY

九里香為蝴蝶，如玉帶鳳蝶提供不可或缺的食物來源和棲息處。

The tree serves as an essential food source and habitat to butterflies such as *Papilio polytes* (Common Mormon).

楊桃 又稱：五斂子、洋桃
Carambola, Star Fruit | *Averrhoa carambola* L.

相片拍攝地點：香港動植物公園
Tree Location: Hong Kong Zoological and Botanical Gardens

名字由來 MEANINGS OF NAME

Carambola 來自梵語 *karmaranga*，意思為開胃菜。而因為其果實的橫切面呈星芒狀，故亦被稱為「星梨」。因以前楊桃是越洋運輸到中國，加上其果實掛在枝條上的姿態，故在漢語中楊桃最初被命名為「洋桃」。「楊桃」是現今較為普及的名稱。

"Carambola" comes from the Sanskrit word *karmaranga*, which means appetizer. Because the cross-section of its fruit is star-shaped, it is also called "Star Fruit". As the carambola was transported across the ocean to China in the past, and its fruit was hung on the branches, so the carambola was originally named *Yangtiao* in Chinese. 「楊桃」is a more popular name in Chinese today.

本地分佈狀態 DISTRIBUTIONS	外來物種 Exotic species
原產地 ORIGIN	馬來西亞、印尼。 Malaysia and Indonesia.
生長習性 GROWING HABIT	常綠喬木。高度可達 8 米。 Evergreen tree. Up to 8 m tall.

1	2	3	4	5	6	7	8	9	10	11	12

花果期月份

花期：本港五月至八月。果期：本港九月至十二月。
Flowering period: May to August in Hong Kong. Fruiting period: September to December in Hong Kong.

　　楊桃是一種聞名於世的果樹，在馬來西亞、泰國和中國等熱帶及亞熱帶地區廣泛種植。楊桃果實形狀獨特，味道酸甜可口，一試難忘，通常在沙拉和其他甜點中用作擺盤裝飾。在中醫學中，楊桃果實有清熱、利尿、解毒作用。然而，過量攝入楊桃可對人類健康產生負面影響，例如引發噁心、失眠、打嗝甚至死亡。因此，還是建議大家「少吃多滋味」。

　　Carambola is widely cultivated for its sweet and juicy fruits in tropical and subtropical regions such as Malaysia, Thailand and China. Other than eating it fresh, the fruit is added for embellishing salads and other desserts in view of its good taste and particular star shape. In traditional Chinese medicine, the fruits are effective in expelling heat, inducing urination and detoxifying. However, excessive ingestion of Carambola can lead to negative health impact, like triggering nausea, insomnia, hiccups or death. and thus should be avoid.

辨認特徵 TRAITS FOR IDENTIFICATION

① 樹幹 TRUNK	② 樹皮 BARK	③ 葉 LEAVES
④ 花 FLOWERS	⑤ 果 FRUITS	

① 楊桃的樹幹。

Trunk of *Averrhoa carambola* L.

② 樹皮暗褐色，光滑，有時具裂縫。

Bark dark brown, smooth, sometimes rimous.

③ 葉片互生，奇數羽狀複葉，小葉 5-11 對，互生至近對生。葉片卵形至橢圓形，頂端漸尖，一側基部歪形。葉面無毛，葉柄和葉軸被柔毛。末端葉片最大。

Imparipinnately compound alternate, leaflets 5-11 pairs alternate to subopposite. Leaflet blade glabrous, ovate to elliptic, base oblique, apex shortly acuminate. Leaf surface glabrous, petiole and leaf rachis pilose. Terminal leaflets always the largest.

④ 圓錐花序於樹幹及樹枝，呈粉紅色。花較小，花瓣 5 片，呈粉紅色。

Inflorescences on trunk and branches, pink. Flowers small, petals 5, pink.

⑤ 漿果，呈黃色至黃棕色，常有 5 棱凸起，橫切面星芒狀。

Berries yellow to yellow brown, always deeply 5-ribbed, stellate in cross section.

生態 ECOLOGY

楊桃被栽培在灌木叢、路邊和花園。

It is cultivated in bushes, roadsides and gardens.

生命力 VITALITY

楊桃對陰暗具高耐受性，它偏好濕潤土壤。

Carambola is generally shade-hardy but prefers moist soils.

幌傘楓 又稱：火通木

Heteropanax fragrans (Roxb.) Seem.

相片拍攝地點：香港中文大學
Tree Location: The Chinese University of Hong Kong

名字由來 MEANINGS OF NAME

由於幌傘楓開花時會散發着陣陣幽香，故其種加詞取為 *fragrans*（拉丁文意譯為芳香）。

The specific epithet *fragrans* because of its fragrance when it blossoms.

應用 APPLICATION

幌傘楓的樹形優美，樹幹直立，枝葉茂密，故經常被種植為觀賞樹和行道樹。另外，其根皮具有治療燒傷、膿瘡和毒蛇咬傷的療效。其木材亦可用於家具製作。

The tree is always planted as an ornamental and street tree in respect of its handsome tree form with prominently straight trunk and dense foliage. In addition, its root bark can be used to treat burns, boils and snakebites. Its wood can be also used for making furniture.

本地分佈狀態 DISTRIBUTIONS	外來物種 Exotic species
原產地 ORIGIN	原生於印度至中國東南部，亦遍及東南亞國家，例如：泰國、越南、老撾、緬甸等。 From India to Southeast China, also Southeast Asian countries, such as Thailand, Vietnam, Laos and Myanmar.
生長習性 GROWING HABIT	常綠喬木。高度可達 10 米。 Evergreen tree. Up to 10 m tall.

花果期 月份	1	2	3	4	5	6	7	8	9	10	11	12

花期：本港十月至十二月。果期：本港二月至三月。
Flowering period: October to December in Hong Kong. Fruiting period: February to March in Hong Kong.

① 樹幹 TRUNK　　② 樹皮 BARK　　③ 葉 LEAVES
④ 花 FLOWERS　　⑤ 果 FRUITS

① 幌傘楓的樹幹。小枝粗壯，具明顯葉痕，頂端被柔毛。

Trunk of *Heteropanax fragrans* (Roxb.) Seem. Branchlets stout, with many observable leaf scars, apex pubescent.

② 樹皮稍粗糙，黃棕色至灰棕色。

Bark slightly rough, yellowish-brown to grayish brown.

③ 葉序為 3 至 5 回羽狀複葉，簇生於小枝頂端。小葉對生，無毛，紙質，橢圓形至橢圓狀卵形，基部楔形，頂端銳形至漸尖，葉緣全緣，兩面側脈明顯。葉柄粗壯，基部隆起，包覆莖。

3-5 pinnately compound, crowded at the apex of branchlets, leaflets opposite. Blade glabrous, chartaceous, elliptic to elliptic-ovate, base cuneate, apex acute to acuminate, entire, lateral veins distinct on both sides. Petiole stout, swollen at the base, sheathing the stem.

④ 雌雄同體。圓錐花序頂生，密被銹色星狀絨毛。花朵呈淡黃白色，芳香。

Hermaphroditic. Panicles terminal, covered with densely rusty stellate tomentose. Flowers pale yellowish white, fragrant.

⑤ 核果卵球形，具凹槽，無毛，花柱宿存，成熟時轉為黑色。

Drupes ovoid-globose, grooved, glabrous, styles persistent, black at maturity.

野生幌傘楓普遍分佈於山谷、山坡、疏林及陽光充沛、水源充足的環境。在缺光的情況下，樹木的橫向發展會受到限制，並驅使其長成幼長的樹幹以爭奪更多的陽光。幼長的樹幹難以支撐整棵樹木，因此有較高的倒塌風險，尤其是颱風等極端天氣期間，容易受強風吹襲，構成公眾安全隱患，故此種植前應先評估是否適合種植在該處。

The wild *Heteropanax fragrans* is commonly distributed in valleys, hillsides, sparse forests and the environments that can support it with enough water and sunlight. Scarce sunlight can restrict the horizontal expansion of the tree and compel it to grow into a lanky stem to compete for more sunlight. This tree form fails to support the tree well and has a higher risk of tree failure, especially during extreme weather periods like typhoons. The tree is vulnerable to strong wind and poses a risk to the public safety. Therefore, the suitability of the planting location should be evaluated before planting,

幌傘楓與菜豆樹 *Heteropanax fragrans* and *Radermachera sinica* (Asia Bell Tree)：

菜豆樹屬於紫葳科，幌傘楓則屬於五加科。在遺傳學而言，兩者的關係較遠。然而，兩者的形態極度相似，難以辨認，通常只能透過其花朵判斷，此外，葉序是區分兩者的關鍵特徵。菜豆樹為複葉對生，幌傘楓則為複葉互生，而且葉柄伸長，緊連小枝。

Asia Bell Tree belongs to Bignoniaceae which is genetically distant from Araliaceae, which *Heteropanax fragrans* belongs to. However, the two trees are highly similar morphologically and it is difficult to distinguish them. Usually, they are only distinguishable by their flowers. Leaf arrangement is also a diagnostic trait to distinguish two trees. While Asia Bell Tree's compound leaves are arranged oppositely, those of *Heteropanax fragrans* are alternately arranged and the petioles are elongated to clasp the branchlets.

海杜果

Cerbera, Sea Mango | *Cerbera manghas* L.

相片拍攝地點：米埔自然保護區、迪欣湖、大埔海濱公園
Tree Location: Mai Po Nature Reserve, Inspiration Lake, Tai Po Waterfront Park

名字由來 MEANINGS OF NAME

種加詞 *manghas* 源自於其形似杜果的果實形狀及葉序。然而，兩個物種之間的關係沒有想像中親密，海杜果為夾竹桃科，杜果為漆樹科。要分辨兩者，可從樹上採集一片葉片，如果是海杜果，斷口處的傷口會分泌出白色汁液。另外，由於這棵樹與杜果相似，但分佈在海岸線上，因此被命名為「海杜果」。

Since the fruit and leaf arrangement of the tree is morphologically closed to Mango (*Mangifera indica*), it is given with the specific epithet *manghas*. However, the relationship between them is not as close as expected; while Sea Mango is the member of Apocynaceae, Mango is in Anacardiaceae. To identify the trees, you can try to collect a piece of leaf from the tree, and only Sea Mango will secret white sap from its wounded part. On account of its mango-like appearance but often found along coastlines, it is named as "Sea Mango".

本地分佈狀態 DISTRIBUTIONS	原生物種 Native species
原產地 ORIGIN	海杜果分佈於奔巴島、西印度洋和太平洋。它亦原生於香港。 Distributed throughout Pemba Island, the West Indian Ocean and the Pacific Ocean. The tree is native to Hong Kong.
生長習性 GROWING HABIT	小型常綠喬木。高度可達 8 米。 Small evergreen tree. Up to 8 m tall.

1	2	3	4	5	6	7	8	9	10	11	12

花果期月份

花期：本港四月至十一月。果期：本港七月至十二月。
Flowering period: April to November in Hong Kong. Fruiting period: July to December in Hong Kong.

　　海杜果雖然有毒，但它的種仁可加工成中藥——牛心茄。其種仁含有強心作用，通常外用或與其他藥物共同服用。

Despite of its toxicity, the extracts from the seed kernel is valued for its excellent cardiotonic effect and can be processed into a traditional Chinese medicine *Niuxinqie*. It is usually used externally or co-administered with other medicines.

辨認特徵 TRAITS FOR IDENTIFICATION

① 樹幹 TRUNK	② 樹皮 BARK	③ 葉 LEAVES
④ 花 FLOWERS	⑤ 果 FRUITS	

① 海杜果的樹幹。枝條輪生，葉痕明顯。

Trunk of *Cerbera manghas* L. Branches whorled, with conspicuous leaf scars.

② 樹皮灰褐色。

Bark grey-brown.

③ 單葉互生，常以螺旋狀排序簇生於枝條末端。葉片橢圓形至倒卵形，基部楔形，頂端漸尖，兩面皆無毛。

Simple leaves alternate, always clustered spirally at the apex of branchlets. Elliptic to obovate, base cuneate, apex acuminate, glabrous on both surfaces.

④ 聚傘花序，芳香。花瓣 5 片，呈白色，中間呈粉色。

Cymes panicle, aromatic. Petals 5, white, centre in pink.

⑤ 核果球形至卵球形，光滑，外果皮纖維質或木質，常具一顆種子，成熟時由綠色轉為紅色。

Drupes globose to ovoid, smooth, always 1-seeded, exocarp cellulosic or woody, turning from green to red when mature.

生態 ECOLOGY

海杜果還進化了一種特殊機制，使其免受種子傳播者短缺的影響，這是沿海地區植物共同面臨的難題。野生海杜果靠海水散播果實。果實內層含豐富木質纖維，使其更容易在海上漂浮，漂流時間更長。

Sea Mango also develops a specific mechanism to prevent itself from being affected by a shortage of seed dispersers, which is a conundrum shared among plants in coastal areas. Wild Sea Mango relies on water for seed dispersal. Its fibrous inner fruit layer allows the fruit to float in the sea more easily and float for a longer time.

生命力 VITALITY

由於高溫、鹽度波動、週期性的乾濕環境以及不穩定和鬆散的土壤基質，一般植物無法在海邊生存。有趣的是，海杜果和一些植物已經進化出對這種惡劣環境的驚人耐受性。而這些植物被稱為「紅樹林伴侶」。

Plants generally cannot survive in coastal areas where feature high temperature, fluctuating salinity and moisture, and loosen substratum. Sea Mango and a small group of plants have evolved to have great tolerance to these harsh environmental conditions, and they are known as "mangrove companion".

植物趣聞 ANECDOTE ON PLANTS

毒性 Toxicity：

大多數夾竹桃科的植物皆含有汁液，汁液具劇毒且種類繁多。海杜果的汁液含豐富強心甙，此汁液分佈在整棵植株，尤其在果實及種子。意外攝取會導致嘔吐、胃痛、四肢麻木，甚至死亡。由於其果實與杜果相似，在中國、斯里蘭卡和印度常有誤食海杜果中毒的報道。

Most members of Apocynaceae contain sap. The sap is lethal and with different chemical compositions among different species. The sap of Sea Mango, which is found within the whole plant, contains profuse cardiac glycosides and concentrated in fruits and seeds. Accidental ingestion can cause vomiting, stomach pain, limb numbness, or even death. Since the fruits are outwardly similar to mangos, poisoning cases of accidental ingestion are often reported in China, Sri Lanka, and India.

夾竹桃

Oleander, Common Oleander | *Nerium oleander* L.

相片拍攝地點：櫻桃街公園、大水坑
Tree Location: Cherry Street Park, Tai Shui Hang

本地分佈狀態 DISTRIBUTIONS	外來物種 Exotic species
原產地 ORIGIN	夾竹桃原生於地中海和中東地帶。 Native to the Mediterranean region and Middle East.
生長習性 GROWING HABIT	常綠灌木。高度可達 6 米。 Evergreen shrub. Up to 6 m tall.

花果期 月份	1	2	3	4	5	6	7	8	9	10	11	12

花期：本港四月至九月。果期：不詳。
Flowering period: April to September in Hong Kong. Fruiting period: Unknown.

屬名 *Nerium* 源自希臘詞 *nerion*，帶有潮濕的含意，形容夾竹桃偏愛潮濕土壤。種加詞 *oleander* 形容夾竹桃的葉片與橄欖葉十分相似。

The generic name *Nerium* is derived from the Greek word *nerion* (moist), referring to its demand for moist soils. The specific epithet *oleander* describes its leaves akin to the leaves of olive tree.

不同國家皆引入夾竹桃作觀賞用途。在香港，由於其花朵出眾，花期悠長，故常被種植於公園和花園內。連綿不絕的秀麗花朵為城市刻畫一幅醉人的風景。夾竹桃的維護成本不高，較少出現病蟲害，對土壤類型的要求不高，對乾旱和貧瘠的土壤具高耐受性。除了以上種種原因外，夾竹桃對二氧化氮、二氧化硫和空氣微粒等空氣污染物具高耐受性，故被廣泛種植在路邊和高速公路，用於綠化和植物修復。

與其他夾竹桃科植物（例如海杧果）一樣，夾竹桃具劇毒，誤食後可引致非常嚴重的後果。不過，攝取適當的劑量對治療疾病是一大良藥。夾竹桃可治療心臟病、癌症和哮喘等疾病。此外，從夾竹桃花中提取的綠色染料具抗炎作用，紓緩皮膚疾病。其葉片能治療心臟病，根部則可治療癌症、潰瘍和麻風病。

夾竹桃毒性有效驅蟲。其汁液可加工成老鼠藥。其葉片萃取物可用於驅除甘蔗中的蟎及常見於柑橘的潛葉蟲，來減輕作物損失。

Oleander is widely introduced into different countries for ornamental purposes. In Hong Kong, the species is always planted in parks and gardens in view of its exquisite and long blooming. The endless and glamorous blossoms help embellish the city with creating a picturesque and graceful urban landscape. Oleander requires relatively low maintenance cost, with few concerns of pest and diseases problems; it is undemanding on soil types, and highly tolerant to drought and infertile soils. Coupled with its remarkable tolerance to air pollutants, such as nitrogen dioxide, sulphur dioxide and particulate matter, it is broadly planted on roadsides and highways for the sake of greening and phytoremediation.

Comparable with other Apocynaceae plants such as Sea Mango (*Cerbera manghas*), Oleander is extremely toxic and the result of misingestion can be drastic. However, it is medicinally effective to treat diseases, such as cardiac illness, cancer and asthma, if consumed with carefully controlled dosage. In addition, the green dye extracted from its flowers is anti-inflammatory and can attenuate skin diseases. Its leaves are medicinally useful to cure heart diseases while the roots can treat cancer, ulcers and leprosy.

In respect of its toxicity, the plant is often used for expelling pests. Its sap can be processed into rat poisons. The extracts from the leaves are insecticidal and commonly used for expelling sugarcane mites and citrus leafminers, which are devastating to crop yields.

① 樹幹 TRUNK	② 樹皮 BARK	③ 葉 LEAVES
④ 花 FLOWERS	⑤ 果 FRUITS	

① 夾竹桃的樹幹。

Trunk of *Nerium oleander* L.

② 樹皮灰色，光滑。

Bark grey, smooth.

③ 單葉 3-4 片，輪生。葉片狹橢圓形，革質，頂端漸尖或銳形，基部楔形或下延於葉柄上。中脈明顯，背面呈脊狀凸起，葉背具明顯密生網狀脈。

Simple leaves 3-4 arranged in a whorl. Blade narrowly elliptic, leathery, apex acuminate or acute, base cuneate or decurrent, midvein prominent abaxially, net veins dense and observable abaxially.

④ 聚繖花序頂生，具分枝。花朵誘人，花被漏斗狀，花瓣 5 或栽培成重瓣花，顏色多樣，可呈紫紅色、粉紅色、白色、橙紅色或黃色。

Cymes terminal, branched. Flowers funnel-shaped, petals 5 or cultivated into double-flowered, colours many, often purplish red, pink, white, salmon, or yellow.

⑤ 蓇葖果圓柱形，成熟時變乾，裂開。

Follicles cylindric, dry and split when ripe.

厚殼樹

Koda Tree │ *Ehretia acuminata* R. Brown

相片拍攝地點：城門標本林、昂坪
Tree Location: Shing Mun Arboretum, Ngong Ping

本地分佈狀態 DISTRIBUTIONS	原生物種 Native species
原產地 ORIGIN	原生於東南亞至澳洲東部的海拔 100 至 1700 米的森林中。 Native to forests with an elevation of 100 to 1700 m above sea level ranging from East Asia and South Asia to Eastern Australia.
生長習性 GROWING HABIT	常綠喬木。高度可達 15 米。 Evergreen tree. Up to 15 m tall.

1	2	3	4	5	6	7	8	9	10	11	12

花果期 月份

花期：本港三月至四月。果期：本港八月至九月。
Flowering period: March to April in Hong Kong. Fruiting period: August to September in Hong Kong.

名字由來 MEANINGS OF NAME

為紀念德國傑出的植物學家——格奧爾格‧迪奧尼修斯‧艾雷特（1708-1770），屬名特此取其名。他是歷史上一位才華洋溢的植物插畫家。艾雷特以協助卡爾‧林奈描繪其在分類植物時所使用的「性系統」而聞名。因他所繪製的插圖極為精確及具科學性，故其繪圖風格亦被後人尊為「林奈風格」，並廣泛採納在植物插畫中。

The generic name *Ehretia* commemorates Georg Dionysius Ehret (1708-1770), who was an illustrious German botanist and botanical illustrator in history. Ehret was well-known with his exquisite depiction of Carl Linnaeus's published "sexual system", which served to aid the rudimentary taxonomic work. His drawing style on plants from a precise and scientific perspective was later credited to "Linnaean style" and has been widely adopted in botanical illustration.

應用 APPLICATION

厚殼樹很少種植在香港作觀賞或遮蔭用途。然而，其茂密的樹冠與清秀的花朵令人賞心悅目，而且遮蔭效果拔群，故被強烈建議作為原生樹種種植在街道和公園中。此外，其成熟的的果實肉質飽滿，能吸引雀鳥啄食，為棲生市區的動物帶來食物來源。厚殼樹亦具藥用價值。其葉片有抗炎和抗糖尿病作用，在中國和印度作為傳統藥物以紓緩急性痢疾，其樹皮亦可加工成果汁以治療發燒和舌痛。

Koda Tree is rarely planted in Hong Kong for ornamental and shading purposes. However, it is highly recommended to be planted as a native alternative to ornamental trees currently planted in streets and parks in Hong Kong, by virtue of its dense foliage and graceful flowers. Likewise, the mature fruits are fleshy and can serve as a delicacy to urban animals. Koda Tree is also valued for its versatile medicinal effects. The leaves, with anti-inflammatory and antidiabetic effects, are commonly used as traditional medicines in China and India for treating acute dysentery. The bark can be processed into juice which is effective for relieving fever and sores on tongues.

辨認特徵 TRAITS FOR IDENTIFICATION

① 樹幹 TRUNK	② 樹皮 BARK	③ 葉 LEAVES
④ 花 FLOWERS		

① 厚殼樹的樹幹。

Trunk of *Ehretia acuminata* R. Brown.

② 樹皮黑灰色。小枝光滑，淺褐色，無毛，具皮孔。

Bark dark greyish. Branchlets smooth, light brown, glabrous, lenticellate.

③ 單葉互生。葉片革質，橢圓形至倒卵形或長橢圓狀倒卵形，基部寬楔形，頂端銳形，葉緣具鋸齒，無毛，葉背沿葉脈被短柔毛。

Simple leaves alternate. Blade coriaceous, elliptic to obovate or oblong-obovate, base broadly cuneate, apex acute, margin serrate, glabrous, pubescent along veins abaxially.

④ 圓錐花序被短柔毛或漸變無毛，頂生。花朵密集，具芳香，花萼杯形，花冠漏斗形，具 5 淺裂，呈白色，雄蕊外露。

Panicles terminal, pubescent or glabrescent. Flowers crowded, fragrant, calyx campanulate, corolla funnel-like, 5-lobed, white, stamens exserted.

註：本樹另有核果，核果球形，成熟時轉為橙黃色，內果皮具皺紋，具 2 粒分核。

Remarks: Drupes globose, turning yellowish orange when mature, endocarp wrinkled, pyrenes 2.

備註 Remarks

本樹木學名根據中國植物誌網頁：

Scientific name of this tree is based on Flora of China：

http://www.efloras.org

台灣女貞 又稱：日本女貞、琉球女貞
Formosa Privet | *Ligustrum liukiuense* Koidz.

相片拍攝地點：大帽山
Tree Location: Tai Mo Shan

名字由來 MEANINGS OF NAME

台灣女貞原生於琉球群島，故其種加詞為 *liukiuense*。

The specific epithet *liukiuense* denotes the tree is originated from the Ryukyu Islands.

本地分佈狀態 DISTRIBUTIONS	原生物種 Native species
原產地 ORIGIN	台灣女貞原生於琉球群島、中國大陸、台灣和香港地區海拔 1000 米至 3000 米的山脈和森林。 Native to the mountains and forests at 1000 m to 3000 m above sea level in Ryukyu Islands, China's mainland, Taiwan and Hong Kong.
生長習性 GROWING HABIT	常綠灌木或小喬木。高度可達 5 米。 Evergreen shrub or small tree. Up to 5 m tall.

花果期 月份	1	2	3	4	5	6	7	8	9	10	11	12

花期：本港三月至六月。果期：本港七月至十二月。
Flowering period: March to June in Hong Kong. Fruiting period: July to December in Hong Kong.

　　女貞屬適合塑型，並對空氣污染具高耐受性。與山指甲相較，台灣女貞較少被種植於花園、公園或路邊作觀賞或植物修復等用途。台灣女貞的用途五花八門，其含有豐富的苯丙烷和黃酮類化合物等抗氧化含物，具清熱止瀉之效。將其種子經翻炒後可加工成為茶。值得一提的是，服用過量台灣女貞可致命，故在加工成食品和藥品時應諮詢專業人士意見。如誤食新鮮果實會傷害內臟黏膜，甚至在服用後 48 至 72 小時內死亡。

　　Ligustrum spp. shows excellent resilience to pruning and air pollution. By virtue of these characteristics, Chinese Privet (*L. sinense*) is pervasively cultivated in parks and streets as solitary trees and fences, serving ornamental and phytoremediation purposes. In the contrary, Formosa Privet is just sporadically planted in parks as a potential alternative of Chinese Privet in respect of its native status to Hong Kong.

　　Formosa Privet is valued for a host of functions. It has profuse antioxidant components such as phenylpropanoids and flavonoids which can effectively clear heat and attenuate diarrhea. The seeds can be processed into tea after drying. Notably, the plant is fundamentally lethal and should be cautiously handled while processing it into food and medicine. Accidental ingestion of the fresh fruits can damage mucous membranes of organs and may cause death after 2 to 3 days.

辨認特徵 TRAITS FOR IDENTIFICATION

① 樹幹 TRUNK	② 樹皮 BARK	③ 葉 LEAVES
④ 花 FLOWERS	⑤ 果 FRUITS	

① 台灣女貞的樹幹。
Trunk of *Ligustrum liukiuense* Koidz.

② 小枝圓柱狀，幼時被微柔毛，漸變無毛。
Branchlets terete, first puberulent, glabrescent.

③ 單葉對生。葉柄具凹槽。葉片革質，無毛，卵形，基部鈍形至銳形急尖。
Simple leaves opposite. Petioles grooved. Blade leathery, glabrous, ovate, base obtuse to acute.

④ 兩性花圓錐花序,頂生。花朵近無梗,呈白色,兩條雄蕊,外露。

Flowers bisexual. Panicles terminal. Flowers subsessile, white, stamens two, protruded.

⑤ 橢球形,細小,成熟時由淡綠色轉為紫色。

Ellipsoid, small, turning pale green to purple at maturity.

植物趣聞 ANECDOTE ON PLANTS

分類學迷思 Taxonomic confusion：

　　台灣女貞因其由地理變異所致的形態不一,其分類狀態一直備受科學家們所爭議。*L. liukiuense* 已取代 *L. japonicum* subsp. *pubescens* 作為目前廣受接納的新種名。台灣女貞在中國曾被命名為 *L. amamianum*,目前已更新為 *L. japonicum*。在某種意義上,*L. japonicum* 應被認為是與 *L. liukiuense* 不同的物種。

　　The taxonomic status of Formosa Privet has been vague with its inconsistent morphology across geographic variations. Currently, *L. liukiuense* is the most accepted scientific name, noted as the rectified name of the published *L. japonicum* subsp. *pubescens*. Formosa Privet in China refers to *L. amamianum* that is currently renamed into *L. japonicum*. In a sense, since two names are accepted scientifically, *L. japonicum* should then be treated as a distinct species from *L. liukiuense*.

山指甲 又稱：小蠟樹
Chinese Privet, Small-leaf Privet | *Ligustrum sinense* Lour.

相片拍攝地點：柴灣公園、川龍
Tree Location: Chai Wan Park, Chuen Lung

名字由來 MEANINGS OF NAME

種加詞指它起源於中國。由於其葉片較女貞細小，故山指甲被俗稱為「小蠟樹」。

The specific epithet *sinense* refers to its origin from China. The common name "Small-leaf Privet" highlights its rather smaller leaf than *Ligustrum lucidum*.

本地分佈狀態 DISTRIBUTIONS	外來物種 Exotic species
原產地 ORIGIN	除中國中部、東南、南部各省份外，山指甲亦分佈在台灣、越南等地區。 The provinces of Central, Southeast and South China, Taiwan; Vietnam.
生長習性 GROWING HABIT	落葉灌木或小喬木。高度可達 7 米。 Deciduous shrub or small tree. Up to 7 m tall.

1	2	3	4	5	6	7	8	9	10	11	12	花果期 月份

花期：本港三月至六月。果期：本港九月至十二月。
Flowering period: March to June in Hong Kong. Fruiting period: September to December in Hong Kong.

　　山指甲的樹皮和葉片具藥用價值，用來根治利尿、解熱、消腫等病徵。由於其芳香的白色花朵在花期可形成壯麗的景色，故作為觀賞灌木或喬木在城市公園和花園中廣泛種植。這品種生長迅速，常被修剪成樹籬或造景植物。

The bark and leaves are primary ingredients of the traditional Chinese medicines for promoting diuresis and relieving heat and swelling. In addition, Chinese Privet is a rather versatile greening component in parks and roadsides. Other than planting in solitary, the tree is fast-growing and malleable, thereby always shaped into multiform hedges and topiaries. Proper pruning remains critical to maintain a decent shape.

辨認特徵 TRAITS FOR IDENTIFICATION

① 樹幹 TRUNK	② 樹皮 BARK	③ 葉 LEAVES
④ 花 FLOWERS	⑤ 果 FRUITS	

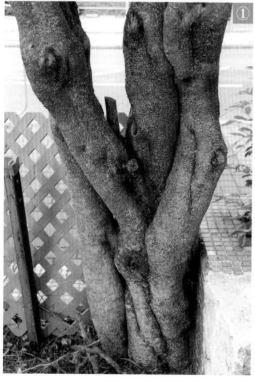

① 山指甲的樹幹。

Trunk of *Ligustrum sinense* Lour.

② 樹皮光滑，灰白色。小枝圓柱狀，常被黃色柔毛。

Bark greyish white, smooth. Branchlets terete, villous to glabrescent.

③ 葉片對生，葉形多變，卵形至披針形，基部寬楔形至近圓形，頂端銳形至鈍形，側脈 4-8 對，無毛至密被絨毛，紙質至薄革質。

Simple leaves opposite. Blade papery to thinly leathery, shapes highly variable, ovate to lanceolate, base broadly cuneate to subrounded, apex acute to obtuse, lateral veins 4-8, impressed adaxially, midvein slightly villous or puberulous.

④ 圓錐花序頂生或腋生，雙性花。花朵呈白色，芳香，花梗無毛或被淡黃色柔毛。

Hermaphroditic. Panicles terminal or axillary, peduncle first yellowish puberulent glabrescent. Flowers small, fragrant.

⑤ 核果近球形，成熟時由綠色轉為暗紫色。

Drupes subglobose, turning green to dark purple when mature.

生態 ECOLOGY

在香港，山指甲通常分佈在灌木叢、路邊和郊區。山指甲的花朵香遠益清，其香味深受昆蟲喜愛，而肉質果實則吸引鳥類和其他動物，故有助於豐富當地的生物多樣性。

In Hong Kong, Chinese Privet is usually found along thickets, roadsides and outskirts. It helps constitute the local biodiversity by providing abundant fragrant flowers and succulent fruits that are appealing to urban animals.

生命力 VITALITY

山指甲對陰暗、乾旱、排水不良或貧瘠的土壤等惡劣環境具高耐受性。

Chinese Privet acclimatises to multiple environments, even when there is shady, drought, poor drainage or barren.

植物趣聞 ANECDOTE ON PLANTS

山指甲與女貞屬植物 Chinese Privet and the genus *Ligustrum*：

女貞屬中以山指甲在香港的分佈最廣，常見於市區和鄉郊地區。有別於無毛的女貞，山指甲幾乎全棵皆被柔毛覆蓋。

Chinese Privet is the most dominant species in *Ligustrum*, with a wide distribution in urban and rural areas. Different from the Glossy Privet that has hairless leaves, Chinese Privet is almost entirely covered with hair.

紫花風鈴木

Purple Tabebuia, Pink Trumpet Tree, Red Lapacho | *Handroanthus impetiginosus* (Mart. ex DC.) Mattos

相片拍攝地點：葵芳邨
Tree Location: Kwai Fong Estate

名字由來 MEANINGS OF NAME

種加詞 *impetiginosus* 源自拉丁詞 *impetigo*，帶有皮膚感染的含意，暗指植物對該類疾病的療效。

本地分佈狀態 DISTRIBUTIONS	外來物種 Exotic species
原產地 ORIGIN	原生於中美洲（墨西哥西北部）至南美洲（阿根廷西北部）。 Ranging from Northwest Mexico to Northwest Argentina.
生長習性 GROWING HABIT	落葉喬木。高度可達 30 米。 Deciduous tree. Up to 30 m tall.

花果期
月份

1	2	3	4	5	6	7	8	9	10	11	12

花期：本港十二月至四月。果期：本港三月至七月。
Flowering period: December to April in Hong Kong. Fruiting period: March to July in Hong Kong.

The specific epithet *impetiginosus* is derived from the Latin word *impetigo*, a kind of skin infection, alluding to the plant's drug effect on the disease.

紫花風鈴木開花時花團錦簇，好像一個個繡球掛滿枝條，極具觀賞價值，故與其他風鈴木一樣被視為觀賞樹種廣泛種植。每逢春天，風鈴木都會綻放出壯麗的花朵，用連綿不絕、色彩繽紛的花朵籠罩着整座城市，回應着人們對希望和新生的嚮往。

除觀賞用途外，紫花風鈴木在原生地中廣泛地用作醫學用途。其樹皮可治療皮膚炎。其葉片可治療外傷，如背痛和牙痛，以及有抗炎作用。巴西人更會用紫花風鈴木的葉片治療蛇毒。此外，紫花風鈴木的提取物蘊含豐富的黃鐘花醌，有抗癌之效。

The tree is introduced into Hong Kong primarily for ornamental purposes, along with other Trumpet trees. Every spring, the Trumpet trees stage a spectacular blooming display and blanket the city in endless flamboyant blossoms, answering humans' the graceful yearn of hope and renewal.

Other than ornamental purposes, Purple Tabebuia is also valued for its versatile medicinal functions in its native range. Its bark can treat skin inflammatory diseases; the leaves can cure physical illness like backache and toothache as they are anti-inflammatory and used by Brazilians for treating snake venom. The bark extracts contain plenty of lapachol that is anti-cancer.

① 樹幹 TRUNK	② 樹皮 BARK	③ 葉 LEAVES
④ 花 FLOWERS	⑤ 果 FRUITS	

③ 掌狀複葉對生，小葉 5-7 片掌狀排列。葉柄兩側近端具葉枕。小葉片卵形至橢圓形，葉緣具鋸齒。兩面小葉較小，中間小葉片較大，其他兩邊小葉末端逐漸向中間收窄。兩側具鱗片，葉背葉脈的葉腋被短柔毛。

① 紫花風鈴木的樹幹。

Trunk of *Handroanthus impetiginosus* (Mart. ex DC.) Mattos.

② 樹皮灰棕色，表面具縱狀皺紋。

Bark greyish brown, longitudinally furrowed.

Palmately compound opposite, leaflets 5-7. Pulvinus at two proximal ends of petioles. Blade ovate to elliptic, serrate, pubescent at the axils of abaxial veins, central leaflet always the largest, lateral leaflets getting smaller.

④ 雌雄同體。花朵簇生於小枝頂端，被星狀毛。花萼鐘狀至管狀，5 齒牙狀。花冠 5 裂，喇叭狀，呈紫粉紅色，花冠喉呈乳黃色。

Hermaphroditic. Flowers clustered at the apex of branchlets, pubescent. Calyx campanulate to tubular, 5-dentated, corolla 5-lobed, trumpet-like, purplish pink, corolla throat cream.

⑤ 蒴果長圓柱形，無毛，呈暗褐色，成熟時縱裂成兩半。種子密集，具一對膜質翅。

Capsules elongated-cylindrical, glabrous , dark brown and longitudinally split into half when mature. Seeds compressed, with a pair of membranous wings.

生命力 VITALITY

　　紫花風鈴木需全日照，對乾旱具高耐受性。排水良好的土壤有利於紫花風鈴木茁壯成長。

Purple Tabebuia shows notable resilience to drought. Good planting requires well-drained soils and full sunlight exposure.

植物趣聞 ANECDOTE ON PLANTS

鐘花樹屬與風鈴木屬 *Tabebuia* and *Handroanthus*：

　　風鈴木泛指鐘花樹屬與風鈴木屬的樹種，兩個屬皆擁有喇叭狀的花朵和掌狀複葉。要分辨兩者，可從花萼的外觀入手。鐘花樹屬的花萼無毛，呈雙唇形，而風鈴木屬的花萼則常被毛狀體，葉緣具 5 齒牙狀。根據分類學系統，*Tabebuia impetiginosa* 已被修改為 *Handroanthus impetiginosa*。

Generally, Trumpet tree refers to the trees from *Tabebuia* and *Handroanthus*, which share trumpet-like flowers and palmate leaves. Two genera, however, are diagnostic from calyx appearance in which *Tabebuia*'s calyx is bilabiate, without lepidote indumentum, while the one of *Handroanthus* is hairy and always 5-dentated at the apex of margin. Based on the updated taxonomic key to the genera, *T. impetiginosa* has been revised into a more proper scientific name *H. impetiginosus*.

備註 Remarks
本樹木學名根據世界植物線上網頁：
Scientific name of this tree is based on Plants of the World Online website:
https://powo.science.kew.org

藍花楹

Jacaranda | *Jacaranda mimosifolia* D. Don

相片拍攝地點：沙田公園、星街
Tree Location: Sha Tin Park, Star Street

名字由來 MEANINGS OF NAME

　　屬名 *Jacaranda* 源自南美圖皮語中的 *jakara'na* 一詞，用以描述其花香沁人心脾。種加詞 *mimosifolia* 則形容其葉片外形與含羞草相似。

The genus name *Jacaranda* is derived from the word *jakara'na* which came from the language of South American Tupi, to illustrate the fragrance of the flowers. The species name *mimosifolia* describes its leaves similar to those of *Mimosa*.

本地分佈狀態 DISTRIBUTIONS	外來物種 Exotic species
原產地 ORIGIN	南美洲，包括巴西、玻利維亞和阿根廷。 In South America, including Brazil, Bolivia and Argentina.
生長習性 GROWING HABIT	常綠喬木。高度可達 15 米。 Evergreen tree. Up to 15 m tall.

1	2	3	4	5	6	7	8	9	10	11	12

花果期月份

花期：本港四月至六月。花期：本港五月至十月。
Flowering period: April to June in Hong Kong. Fruiting period: May to October in Hong Kong.

　　藍花楹的木材柔軟，可作為製作家具的原材料，故是不可或缺的木材樹種。此外，藍花楹具高觀賞價值。其盈盈的薰衣草色花朵搖曳生姿，構成使人心醉的景色。南非比勒陀利亞被譽為「藍花楹之城」，在花期，無窮無盡的紫色花朵綻放，為這城市增添神秘感。

Jacaranda is an important timber species. The wood is soft and can be used as a raw material for making furniture. Besides, the tree has high ornamental value with attractive lavender colour flowers. It creates a spectacular view when the flowers are proliferated. Pretoria in South Africa is regarded as the "City of Jacarandas". During the flower season, the city is immersed in purple with endless blooms.

辨認特徵 TRAITS FOR IDENTIFICATION

① 樹幹 TRUNK	② 樹皮 BARK	③ 葉 LEAVES
④ 花 FLOWERS	⑤ 果 FRUITS	

① 藍花楹的樹幹。

Trunk of *Jacaranda mimosifolia* D. Don.

② 樹皮褐色，縱裂。

Bark brown, with longitudinal cracks.

③ 二回羽狀複葉對生，葉片多於 10 對，對生，小葉 16-24 對，奇數羽狀複葉對生，無柄，頂端葉片較大，葉軸具窄翅。

Oppositely bipinnate, pinnae more than 10 pairs, opposite, leaflets 16–24 pairs, imparipinnate, opposite, sessile, with larger terminal leaflet, rachis with narrow wings.

④ 圓錐花序頂生。花冠呈藍紫色，漏斗狀，被短柔毛。

Panicles terminal. Corolla bluish purple, funnelform, pubescent.

⑤ 蒴果，木質，扁圓形，葉緣常捲曲，成熟時由綠色轉為暗褐色。

Capsules woody, oblate, margin always curvy, turning green to dark brown when mature.

生態 ECOLOGY

藍花楹已經受到嚴重的砍伐活動的威脅。2019 年 IUCN 紅色名錄中，它被列為「易危」。該樹種被引進到其他國家作為觀賞用途，但在南非被認為是入侵品種。藍花楹可忍受晴朗天氣和乾燥土壤。其花受到眾多傳粉者的喜愛，這可能會促進該樹在熱帶地區的歸化和傳播。

The wild Jacaranda has been threatened by the drastic logging activity. It has been categorized as Vulnerable in the IUCN Red List of Threatened Species in 2019. The tree has been introduced to other countries for ornamental purposes; however, has been noticed as an intruder in South Africa. Jacaranda can tolerate sunny weather and dry soils. Its flowers are beloved by numerous pollinators, this may promote the naturalization and dissemination of the tree in tropical regions.

植物趣聞 ANECDOTE ON PLANTS

藍花楹與鳳凰木的分別 Jacaranda and *Delonix regia* (Flame Tree)：

雖然鳳凰木別名「紅花楹」，與藍花楹的名稱相似，但兩者屬於不同科屬，藍花楹為紫葳科，而鳳凰木則為蘇木科。要分辨兩者，可從花色、樹皮、葉序及果實入手。藍花楹的花朵為紫藍色，鳳凰木則呈紅色。藍花楹的樹皮縱裂，而鳳凰木則較粗糙，不具裂痕。藍花楹擁有對生葉序，而鳳凰木則為互生。藍花楹的果實為蒴果，而鳳凰木則為豆莢狀。

In Chinese, Flame Tree (*Delonix regia*) is also known as "Red-flower Pillar", whereas Jacaranda, is commonly named as "Blue-flower Pillar". Nevertheless, they are indeed referred to different families. Jacaranda is under Bignoniaceae while Flame Tree is under Caesalpiniaceae. They can be concisely distinguished when blossoming, with Jacaranda purplish blue and Flame Tree red. Besides, the bark of Jacaranda is longitudinal cracked, while that of Flame Tree is slightly coarse, without any longitudinal cracks. The pinnae of Jacaranda are oppositely arranged, while that of Flame Tree are alternately arranged. About the type of fruit, the fruit of Jacaranda is capsule, and those of Flame Tree are legumes.

吊瓜樹 又稱：吊燈樹

Sausage Tree, Cucumber Tree | *Kigelia africana* (Lam.) Benth.

相片拍攝地點：香港大會堂、新興花園
Tree Location: Hong Kong City Hall, Sun Hing Garden

名字由來 MEANINGS OF NAME

屬名 *Kigelia* 源自莫桑比克語中的 *kigeli keia*。種加詞 *africana* 意指本樹種的原產地非洲。由於樹上懸掛着圓柱狀、碩大的果實，故有「吊瓜樹」此俗稱。

The generic name *Kigelia* is derived from *kigeli keia*, the Mozambique vernacular of how the natives refer to this genus. The specific epithet *africana* means "from Africa". In respect of its giant sausage-like ellipsoid fruits unexpectedly hanging on the branches, the tree is thus given the names "Sausage Tree" and「吊瓜樹」.

應用 APPLICATION

吊瓜樹的果實具藥效，可用於緩解消化系統疾病。種子烘烤後可食用。在非洲，吊瓜樹的果實被視為祭品，用作祈求豐收、繁殖力、財富和繁榮。

本地分佈狀態 DISTRIBUTIONS	外來物種 Exotic species
原產地 ORIGIN	非洲國家，如埃塞俄比亞、尼日爾和烏干達。 Countries of Africa, such as Ethiopia, Niger and Uganda.
生長習性 GROWING HABIT	常綠喬木。高度可達 25 米 Evergreen tree. Up to 25 m tall.

花果期 月份	1	2	3	4	5	6	7	8	9	10	11	12

花期：本港三月至五月。果期：本港六月至八月。
Flowering period: March to May in Hong Kong. Fruiting period: June to August in Hong Kong.

吊瓜樹古樹名木

　　吊瓜樹因其樹冠闊大、顯眼花朵、獨特果形而被引入作為行道樹或觀賞樹種。此品種果實很重，通常可達至 10 公斤，並可懸掛在樹上一段時間。因此，種植此品種時應謹慎考慮，以免其果實對人命財產構成潛在風險。在大會堂紀念花園有一棵吊瓜樹被登記為古樹名木（編號：LCSD CW/124），其胸徑為 850 毫米，高度為 14 米，樹冠 16 米。大家參觀時緊記小心保護頭部，以免被果實砸傷！

The fruits are medicinally effective for relieving digestive system disorders. The seeds are esculent and always baked for food by the locals. In Africa, the fruits are common charms for praying better crop yields, fecundity, wealth and prosperity.

Sausage Tree is introduced as roadside or ornamental trees by virtue of its extensive tree crown, flamboyant flowers and majestic sausage-like fruits. The fruits are weighty (to 10 kg) but can incredibly cling to the branches for a long time. However, they are still dangerous from which the fruits falling off can be dramatically violent. The planting framework of the tree should be carefully established to avert compromising the safety of human property. In Hong Kong, a Sausage Tree is registered as an old and valuable tree (OVT) in the City Hall Memorial Garden (LCSD CW/124), measured with a DBH of 850 mm, a height of 14 m and a crown spread of 16 m.

辨認特徵 TRAITS FOR IDENTIFICATION

① 樹幹 TRUNK	② 樹皮 BARK	③ 葉 LEAVES
④ 花 FLOWERS	⑤ 果 FRUITS	

① 吊瓜樹的樹幹。
Trunk of *Kigelia africana* (Lam.) Benth.

② 樹皮粗糙，灰棕色，薄片狀。
Bark rough, greyish brown, thinly flaky.

③ 奇數羽狀複葉，對生或輪生 3-4 片，小葉 7-9 對對生。小葉葉片橢圓形至長橢圓形，頂端突尖，基部楔形，葉緣全緣，側脈明顯，無毛。

Imparipinnately compound, opposite or 3-4 in a whorl, leaflets 7-9 pairs, opposite. Blade glabrous, elliptic to oblong, apex cuspidate, base cuneate, entire, lateral veins predominant.

④ 6-10 朵花朵簇生成圓錐花序，花序着生於枝條末端，下垂。大型花朵，呈紅色，花萼鐘狀，花冠漏斗狀，夜間開花。

Panicles terminal at the end of branches, pendulous, flowers 6-10. Flowers large, scarlet, calyx campanulate, corolla funnelform, blooming in the evening.

⑤ 蒴果呈香腸狀，重，果實閉合不裂，成熟時變為木質化，轉為棕黃色，眾多種子。

Capsules sausage-shaped, heavy, fruit closed indehiscent, woody, brownish yellow at maturity, seeds many.

生態 ECOLOGY

像動物一樣，植物也有夜行性的。以吊瓜樹為例，它與蝙蝠之間具互利共生的關係。不少蝙蝠品種，例如果蝠，為夜行性動物，只在夜間活躍，它們是體型較大的哺乳類動物，雖然視力欠佳，但嗅覺靈敏。為了吸引蝙蝠幫忙授粉，吊瓜樹發展出帶有化學信號（如乙酸異戊酯）的花蜜，讓蝙蝠探測到這些信號後幫忙授粉。

此品種花朵已經進化成擁有更大的花筒和堅固的花梗以承受蝙蝠的體型，容許蝙蝠輕而易舉地深入採蜜。花朵在夜間盛放，開花時間短，通常在翌日下午便會凋落。儘管鳥類亦是傳粉者，但吊瓜樹花開花落與鳥類的作息時間錯開，而其他夜行性動物的體型不如蝙蝠，所以蝙蝠仍是此物種最優秀的傳粉媒介。吊瓜樹的果實累累、沉甸甸的，沒有大型動物的幫助下，果實只能靠重力近距離散播在與母樹附近。

Blossoming is not only confined to daytime. For plants which blossom at night, it is pivotal for them to equip alternative reproductive strategies for aligning with the behaviours of nocturnal pollinators. Here, Sausage Tree has developed an intriguing mutualistic relationship with bats. Bat species (e.g. fruit bat) are only active at night; they are sizable mammals with generally reduced visual acuity but with compensated superb olfactory acuity. Floral colour could be trivial to highlight flowers to bats; therefore, Sausage Tree has evolved brilliantly to produce nectar with chemical signals (e.g. isoamyl acetate) which can be deciphered by bats.

The flowers have developed into larger mouths and stiffer pedicles which can hold a bat and allow it to access the inside nectar. The flowers blossom at night with short duration, often falling off before next afternoon. Although birds could be potential pollinators, they are far less effective than bats and other nocturnal animals due to the floral abscission. The fruits are large and weighty. Without any aid of large animals (e.g. elephants), the fruits is restricted to gravity, which only confers a close dispersal distance from the maternal tree.

貓尾木 又稱：西南貓尾木

Cat-tail Tree | *Markhamia stipulata* (Wall.) Seem. ex K. Schum. var. *kerrii* Sprague

相片拍攝地點：荔枝角公園
Tree Location: Lai Chi Kok Park

名字由來 MEANINGS OF NAME

由於其果實形狀神似貓尾，故被稱為貓尾木。傳説中，貓居住於天堂並掌管人間。有一天，牠們被差派至人間解決囓齒動物橫行侵襲的問題。起初，貓熱衷於捕捉老鼠，後來大概因為感到疲憊不堪和枯燥乏味，遂與老鼠談判，並達成共識，從此不再捕捉老鼠。貓因而變得怠惰，而且日益心廣體胖，結果導致鼠患問題一發不可收拾。鼠患失控令眾神大為震驚，牠們決定嚴懲這群好吃懶做的肥貓。貓群為了逃離眾神的追捕，紛紛化身為樹，並把尾巴變作果實。這就是中國傳説「貓尾木」的由來。

本地分佈狀態 DISTRIBUTIONS	外來物種 Exotic species
原產地 ORIGIN	廣泛分佈於中國華南地區和東南亞地區，如泰國與老撾。 South China and Southeast Asia, such as Thailand and Laos.
生長習性 GROWING HABIT	常綠喬木。高度可達 15 米。 Evergreen tree. Up to 15 m tall.

1	2	3	4	5	6	7	8	9	10	11	12	花果期月份

花期：本港十月至十一月。果期：本港四月至六月。
Flowering period: October to November in Hong Kong. Fruiting period: April to June in Hong Kong.

The common name "Cat-tail Tree" refers to its cat-tail like fruits. In legend from China, cats lived in heaven and were responsible for human world. In one day, they were sent to solve a rodent infestation. Initially, they were very vigorous of foraging mice. Could because of being fatigued and bored, the cats made a reconciliation with mice and started being apathetic and fat. The rodent problem developed rampantly and ultimately became intractable. The news shocked the gods and they decided to punish the cats for their undutifulness. To evade capture by the gods, the cats transformed themselves into trees while changed their tails into fruits.

應用 APPLICATION

　　貓尾木因其秀美誘人的花朵、獨特有趣的果實，成為公園及花園的常見觀賞性樹種。當你有緣遇見正結果的貓尾木，就感受一下它那鬆軟、毛茸茸的果皮吧！

Cat-tail Tree is always planted for its excellent ornamental value in parks and gardens in respect of its outstanding blossoms and fruits. When you see its fruits next time, enjoy its fluffy pericarp!

辨認特徵 TRAITS FOR IDENTIFICATION

| ① 樹幹 TRUNK | ② 樹皮 BARK | ③ 葉 LEAVES |
| ④ 花 FLOWERS | ⑤ 果 FRUITS | |

① 貓尾木的樹幹。

Trunk of *Markhamia stipulata* (Wall.) Seem. ex K. Schum. var. *kerrii* Sprague.

② 初時密被銹黃色短柔毛。

Initially densely rusty yellow pubescent.

③ 葉為奇數羽狀複葉，小葉 3-9 片對生，單葉於葉柄底部退化。葉片紙質，矩圓形、長橢圓形至矩圓形、長橢圓形至被針形，或卵形至被針形。頂端尾狀至漸尖，基部圓形，葉緣全緣至細鋸齒狀。年幼時密被銹黃色短柔毛，成熟時無毛至稀疏。

Imparipinnately compound opposite, leaflets 3-9 pairs, opposite, subsessile, with a pair of reduced leaflets at the base of petiole. Leaflet blade papery, densely rusty yellow pubescent when young, adaxially glabrescent, abaxially pubescent, oblong, elliptic-oblong, elliptic-lanceolate or ovate-lanceolate, apex caudate-acuminate, base rounded, entire to serrulate.

④ 花瓣 5 片，漏斗狀，呈不規則狀皺摺，黃色，冠筒底部深紫色。雄蕊和花柱藏於花冠內。

5 petals, corolla yellow, tube dark purple, funnelform, lobes irregularly wrinkled. Stamens and style included.

⑤ 蒴果線形，扁平，密被黃褐色絨毛，猶如貓尾。種子長橢圓形，有翅。

Capsules linear, compressed, densely yellowish brown tomentose, like a cat tail. Seeds long elliptic, winged.

生態 ECOLOGY

　　貓尾木在一些國家被當作先鋒樹種，其花朵和果實對鳥類和蝙蝠具不俗的吸引力。

　　Cat-tail Tree is planted as a pioneer species in some countries. Its flowers and fruits are appealing to birds and bats.

銀鱗風鈴木 又稱：黃金風鈴木、銀鱗金鈴木

Silver Trumpet Tree, Tree of Gold | *Tabebuia argentea* (Bureau & K. Schum.) Britton

相片拍攝地點：葵芳邨、大埔中心
Tree Location: Kwai Fong Estate, Tai Po Centre

名字由來 MEANINGS OF NAME

當陽光灑落在銀鱗風鈴木具有光澤的葉片上時，會呈現出閃爍的銀光，像披上了一層薄薄的銀色鱗片，其種加詞 *argentea* 即為「銀」的意思，它亦有 Silver Trumpet Tree 的俗名（中文意譯為銀喇叭樹）。開花時，一簇簇金黃色的花團掛滿枝條，故又有 Tree of Gold（中文意譯為黃金樹）此美稱。

The specific epithet *argentea* refers to silvery, describing its lustrous leaf colour under the sunlight. Due to this characteristic, the tree is named "Silver Trumpet Tee". When blossoms, the crown is entirely covered with golden colour, hence also named as "Tree of Gold".

本地分佈狀態 DISTRIBUTIONS	外來物種 Exotic species
原產地 ORIGIN	中美洲至南美洲。 Mesoamerica to South America.
生長習性 GROWING HABIT	落葉或常綠喬木。高度可達 16 米。 Deciduous or evergreen tree. Up to 16 m tall.

花果期 月份

1	2	3	4	5	6	7	8	9	10	11	12

花期：本港三至四月及九至十月。果期：本港九至十一月。
Flowering period: March to April and September to October in Hong Kong. Fruiting period: September to November in Hong Kong.

　　銀鱗風鈴木開花時花團錦簇，形成一片金燦燦的花海，極具觀賞價值，故與其他風鈴木一樣被視為觀賞樹種廣泛種植。每年春天，各種風鈴木皆在大街小巷中悄悄盛放，爭奇鬥艷，為城市增添活力。路過的行人都不禁在繁忙而喧鬧的都市中佇足細賞，陶醉在這片短暫而壯麗的絕色之中。

　　除觀賞用途外，銀鱗風鈴木在其原生地亦被作為經濟樹木而廣泛種植。此外，銀鱗風鈴木蘊含豐富的類黃酮化合物，當地人會從其葉片和花提取此化合物治療流感。其木材堅硬，有紋理和柔韌性，因此，它通常被用來製造家具和其他建築材料。

In line with other Trumpet trees, Silver Trumpet Tree is also regarded as an excellent ornamental tree due to its spectacular blossoms. Every spring, rows of Trumpet trees blossom and compete for beauty. They immerse the city into a perceived vitality and inevitably slow down every citizen for appreciating this ephemeral but majestic moment.

The tree is also exploited for multiple functions in its native range. The tree contains rich flavonoid compounds. The extracts from leaves and flowers are used for treating influenza. The wood is hard, textured and flexible; therefore, it is commonly harvested for making furniture and other constructions.

辨認特徵 TRAITS FOR IDENTIFICATION

① 樹幹 TRUNK	② 樹皮 BARK	③ 葉 LEAVES
④ 花 FLOWERS	⑤ 果 FRUITS	

① 銀鱗風鈴木的樹幹。

Trunk of *Tabebuia argentea* (Bureau & K. Schum.) Britton.

② 樹皮淡褐色，具明顯縱狀裂紋。

Bark light brown, with cracks and obvious longitudinal lines.

③ 掌狀複葉對生，5-7 片小葉。小葉片呈長橢圓狀橢圓形至長橢圓狀披針形，頂端微凹至圓形，基部近心形至圓形，葉緣全緣，葉背具鱗片狀紋路。陽光直射下呈銀色。

Palmately compound opposite, with 5-7 leaflets. Blade oblong-elliptic to oblong lanceolate, apex retuse to rounded, base subcordate to rounded, entire, abaxially covered with lepidote indumentum. Lustrous when directly exposed to the sunlight.

④ 圓錐花序頂生，具鱗片。花萼鐘狀至二唇形，黃棕色。花冠金黃色，漏斗狀。雌雄同體。

Panicles terminal, covered with lepidote indumentum. Calyx campanulate to bilabiate, yellow brown. Corolla golden, funnelform. Hermaphroditic.

⑤ 蒴果呈長橢圓形，暗褐色，成熟時裂開。種子密集，具翅。

Capsules oblong, dark brown and dehiscent at maturity. Seeds compressed, winged.

植物趣聞 ANECDOTE ON PLANTS

銀鱗風鈴木與黃花風鈴木 Silver Trumpet Tree and *Handroanthus chrysanthus* (Yellow Pui)：

　　銀鱗風鈴木與黃花風鈴木的外形十分相似，兩者皆具有掌狀複葉及金黃色花朵。植株上的毛狀外被能帶給我們一些辨認小貼士！黃花風鈴木的葉被着肉眼觀察得到的星狀毛，銀鱗風鈴木則被着只能透過顯微鏡看見的鱗片。兩者的葉緣亦截然不同，銀鱗風鈴木的葉緣全緣，黃花風鈴木的葉緣則呈鋸齒狀。

Two Trumpet trees are outwardly the same, sharing both palmate compound leaves and golden blossoms. Looking for the types of indumenta covering on plants is the distinguishable trick. The leaves and flowers of Yellow Pui are covered with observable stellate-hairs, while those of Silver Trumpet Tree are lepidote and only can be seen through a microscope. Likewise, whilst the leaves of Yellow Pui are serrate, those of Silver Trumpet Tree are entire.

洋紅風鈴木

Rosy Trumpet Tree, Pink Tecoma │ *Tabebuia rosea* (Bertol.) DC.

相片拍攝地點：詩歌舞街
Tree Location: Sycamore Street

名字由來 MEANINGS OF NAME

其屬名 *Tabebuia* 源自於巴西本土名稱 *tabebuia* 和 *taiaverulia*。洋紅風鈴木的花朵呈玫紅色，故其種加詞為 *rosea*（中文意譯為薔薇）。其擁有喇叭狀的花朵形態，令它亦有 Rosy Trumpet Tree 的俗稱（中文意譯為薔薇喇叭樹）。

The generic name *Tabebuia* comes from *tabebuia* and *taiaverulia*, the Brazilian vernaculars how the local people refer to this genus. The specific epithet *rosea* refers to its rose blossoms. In respect of its trumpet-like blossoms, it is also given the common name as "Rosy Trumpet Tree".

本地分佈狀態 DISTRIBUTIONS	外來物種 Exotic species
原產地 ORIGIN	中美洲至南美洲，包括委內瑞拉、哥倫比亞、厄瓜多爾和法屬圭亞那。 Ranging from Mesoamerica to South America, including Venezuela, Colombia, Ecuador and French Guiana.
生長習性 GROWING HABIT	落葉喬木。高度可達 30 米。 Deciduous tree. Up to 30 m tall.

1	2	3	4	5	6	7	8	9	10	11	12	花果期 月份

花期：本港三至四月。果期：不詳。
Flowering period: March to April in Hong Kong. Fruiting period: Unknown.

　　洋紅風鈴木因其艷麗的花朵被視為觀賞價值高的觀賞樹種。春天來臨時，洋紅風鈴木枝條掛滿盛放的粉嫩花朵，隨着春風擺動，為城市增添嫵媚的氣息。洋紅風鈴木亦是薩爾瓦多的國樹。在當地方言中，此樹被稱為 *Maquilíshuat*。

　　在美洲，洋紅風鈴木具有多種用途。在哥倫比亞北部海岸，當地人經常會利用洋紅風鈴木的樹皮提取物治療由真菌和酵母菌引致的疾病。此外，有研究資料顯示，洋紅風鈴木的甲醇提取物具有抗潰瘍、抗分枝桿菌及保肝作用，可見其具有藥物研發的潛力。在中美洲，洋紅風鈴木的木材常用於家具製作、船舶製造及其他建築材料。由此可見，洋紅風鈴木的用途十分廣泛。

Rosy Trumpet Tree is highly regarded as an ornamental tree mainly due to its spectacular rose blossoms during the spring. As a result, the tree is nominated as the national tree of El Salvador, vernacularly known as *Maquilíshuat.*

The tree is versatile in America for multiple uses. In the Northern Coast of Colombia, people often use its bark extract for tackling the skin diseases induced by fungi and yeast. In addition, some studies reveal that the methanolic extracts of Rosy Trumpet Tree are anti-ulcerogenic, antimycobacterial and hepatoprotective; they are potential for drug development. In Mesoamerica, the wood is demanded for making furniture, ships and other constructions.

辨認特徵 TRAITS FOR IDENTIFICATION

① 樹幹 TRUNK	② 樹皮 BARK	③ 葉 LEAVES
④ 花 FLOWERS	⑤ 果 FRUITS	

① 洋紅風鈴木的樹幹。小枝下垂，具刺。
Trunk of *Tabebuia rosea* (Bertol.) DC. Branchlets drooping with thorns.

② 樹皮淡灰色，堅韌，具縱向皺紋。
Bark greyish, tough with longitudinally furrowed.

③ 掌狀複葉對生，具 5 片小葉。
Palmately compound opposite, leaflets 5.

④ 花萼鐘狀至二唇形，被鱗片毛被覆蓋。
Calyx campanulate to bilabiate, covered with lepidote indumentum.

⑤ 蒴果長圓柱形，暗棕色，成熟時開裂。
Capsules elongated cylindrical, dark brown and dehiscent at maturity.

植物趣聞 ANECDOTE ON PLANTS

洋紅風鈴木與紫花風鈴木 Rosy Trumpet Tree and *Handroanthus impetiginosus* (Purple Tabebuia)：

　　雖然洋紅風鈴木與紫花風鈴木屬於不同屬，但兩者形態十分相似，令人難以辨認。即使開花，兩者皆以喇叭狀、紫粉色的花朵呈現，仿如一對雙胞胎，讓人摸不着頭腦如何區分。毛被是辨認兩者的關鍵特徵。紫花風鈴木的葉脈和花朵通常被柔毛，洋紅風鈴木的葉脈和花朵則具鱗片狀附着器或無毛。與此同時，紫花風鈴木的花萼頂端呈 5 齒牙狀，洋紅風鈴木的花萼頂端則呈二唇狀。

　　Although the trees are categorized into different genus, they are outwardly imperceptible, with showy trumpet-like flowers in purplish pink. The types of indumenta covering on plants are a trick for identifying the trees. While leaf veins and flowers of Purple Tabebuia are covered with hairs, those of Rosy Trumpet Tree are lepidote. Likewise, the calyx of Purple Tabebuia tends to be 5-dentated at the apex, while the one of Rosy Trumpet Tree is bilabiate.

備註 Remarks
本樹木學名根據世界植物線上網頁：
Scientific name of this tree is based on Plants of the World Online website:
https://powo.science.kew.org

檳榔

Betel Palm, Areca Nut, Pinang | *Areca catechu* Linnaeus

相片拍攝地點：九龍公園
Tree Location: Kowloon Park

名字由來 MEANINGS OF NAME

種加詞 *catechu* 源自馬來亞語 *caccu*，意譯為植物萃取液。檳榔的果實可食用，常與食用石灰和蔞葉（又稱檳榔葉）一同共用，加快果實內植物鹼釋放，又名蔞葉棕櫚。

The specific epithet *catechu* is derived from the Malayan word *caccu*, referring to the sap that can be extracted from the plant. Its fruits are edible and always associated with edible lime and betel leaves (*Piper betle*, Betel) to accelerate drug release, hence named as Betel Palm.

本地分佈狀態 DISTRIBUTIONS	外來物種 Exotic species
原產地 ORIGIN	菲律賓。 The Philippines.
生長習性 GROWING HABIT	常綠，單一主莖棕櫚。高度可達 20 米。 Evergreen, solitary palm. Up to 20 m tall.

花果期月份

1	2	3	4	5	6	7	8	9	10	11	12

花期：本港七月至八月。果期：本港十一月至五月。
Flowering period: July to August in Hong Kong. Fruiting period: November to May in Hong Kong.

根據曾到訪過東南亞國家的旅客敘述，當地人常手握一袋檳榔，牙齒因咀嚼檳榔而染上紅色。檳榔是東南亞國家的家傳戶曉的零食。當地人通常把檳榔切成方塊狀，與食用石灰和蔞葉一同食用。咀嚼時，檳榔會釋出單寧及檳榔鹼，別具風味，還能消除疲累。雖然檳榔廣受歡迎，但過度嚼食會引致口腔癌。檳榔內含有檳榔鹼為致幻成分，故在土耳其和澳洲等國家被視為毒品，與檳榔相關的交易受到限制。

檳榔對台灣具極大文化意義，特別是原住民一族。例如，噶瑪蘭族會利用檳榔的汁液把纖維布料染成香蕉色。為祈求繁榮昌盛，西拉雅族人會以檳榔和其他祭物如酒和米飯向他們信奉的神明「阿立祖」獻祭。阿美族視檳榔為信物，男女之間會以檳榔傳情。豐收季第四天後，阿美族姑娘會將一顆檳榔碾碎，並放進心儀男生的檳榔袋中。若果該名男生對那姑娘也有意思，便會把那顆碎掉的檳榔吃掉。故此，檳榔袋在阿美族中被稱為 alofo，即是大家耳熟能詳的「情人袋」。

不過，檳榔在香港主要因為具堅韌的結構和獨特的巨大葉片而作為觀賞植物種植，不作其他用途。

For those who have travelling experiences in Southeast Asian countries, it is not hard to see the locals clasping a bag of palm nuts on their hands with teeth dyed in red. Betel Palm's fruit is a notable snack in Southeast Asian countries. The locals always chop the nuts into cubes and eat with edible lime and betel leaves. When chewing, the released tannin and arecoline from the nuts not only offer people multiple tastes, but also reduce fatigue. Despite its popularity in Asia, it is always bonded with oral cancer when over-consumed. Since the fruits contain abundant arecoline that is hallucinogenic, they are termed as drugs in countries like Turkey and Australia and any trading of the nuts is forbidden.

Betel Palm is culturally weighty to Taiwan, constituting part of the aboriginal cultures. For example, the sap extracted from the palm is used by the Kebalan for dying fabrics into banana colour. The fruits are ritually important as offerings while the Siraya pray to their deity *Ali-zu* for prosperity, coupled with other offerings like wine and rice. The fruits are a medium for expressing affection in the Amis tribe. After four days of harvesting in a year, an Amis girl will put one Betel Palm's nut into the bag of the boy she has a crush on, then the boy will eat it if he shares the same thought as the girl. In respect of the reason, the bag in Amis is named as *alofo*, referring to "valentine bag".

In Hong Kong, however, the application of the palm is limited, barely served as an ornamental palm by virtue of its erect stem and remarkable leaf size.

檳榔偏愛炎熱、全日照、肥沃、排水良好的土壤。

Betel Palm prefers high-temperature, full sunlight, with fertile and well-hydrated soils.

備註 Remarks

本樹木學名根據中國植物誌網頁：

Scientific name of this tree is based on Flora of China :

http://www.efloras.org

① 樹幹 TRUNK	② 樹皮 BARK	③ 葉 LEAVES
④ 花 FLOWERS	⑤ 果 FRUITS	

① 檳榔的樹幹。

Trunk of *Areca catechu* Linnaeus.

② 莖樹狀，直立，灰色，具明顯葉痕。

Stem arborescent, erect, grey, with notable leaf scars.

③ 葉鞘重疊成冠軸，呈綠色。羽狀複葉，大型，每側具 20-30 片小葉。小葉披針形，頂端具歪齒，頂生小葉合生成魚尾狀。

Leaf sheaths overlapping into a crownshaft, green. Pinnately compound, large, with 20-30 leaflets per side. Blade lanceolate, apex obliquely toothed, terminal leaflet merged into fishtail-like shape.

④ 雌雄同株。圓錐花序着生於冠軸基部，分枝為 3 條，直立，穗軸眾多。雄花互生，於穗軸上排成二列，花瓣 3 片，呈黃白色，芳香。雌花大型，着生於穗軸基部。

Monoecious. Panicles growing at the base of crownshaft, branched to 3 orders, erect, rachillae many. Male flowers alternate and distichous on rachillae, petals 3, yellowish-white, fragrant. Female flowers large, only at the base of rachillae.

⑤ 卵形或橢圓形，堅硬，具光澤，果實內部具纖維，成熟時轉為橙黃色。

Drupes ovoid or ellipsoid, hard, lustrous, fibrous inside, turning yellow orange when ripe.

三藥檳榔

Triandra Palm, Wild Areca Palm | *Areca triandra* Roxb. ex Buch. -Ham.

相片拍攝地點：青衣公園、香港公園
Tree Location: Tsing Yi Park, Hong Kong Park

名字由來 MEANINGS OF NAME

馬拉巴爾海岸的當地人對此屬植物的稱為 *areca*，故其屬名名為 *Areca*。種加詞 *triandra* 意指其雄花具三條雄蕊。

The generic name *Areca* is the vernacular how the local people on the Malabar Coast refer to this genus. The specific epithet *triandra* describes the male flowers showing three stamens.

應用 APPLICATION

在眾多檳榔屬種中，三藥檳榔為少數的叢生棕櫚，故通常被視為自然界中的藝術品種植在公園內。種植三藥檳榔能輕易形成一排排天然樹籬以劃分區域，能提升空間的隱密度。

其果實可食用，並能治療由消化不良或腹瀉引起的腹痛。

本地分佈狀態 DISTRIBUTIONS	外來物種 Exotic species
原產地 ORIGIN	東南亞內陸。 Mainland Southeast Asia.
生長習性 GROWING HABIT	常綠，叢生棕櫚。高度可達 5 米。 Evergreen, clustered palm. Up to 5 m tall.

1	2	3	4	5	6	7	8	9	10	11	12	花果期 月份

花期：不詳。果期：不詳。
Flowering period: Unknown. Fruiting period: Unknown.

In view of its clump habit that is occasionally characterised by other *Areca* spp., it is always planted as a visual alternative for parks. By aligning the palm into a row, it can also create a magnificent fencing effect for privacy and safety purposes.

The fruits are edible and always used for treating abdominal pain caused by food accumulation and diarrhea.

① 樹幹 TRUNK	② 樹皮 BARK	③ 葉 LEAVES
④ 花 FLOWERS	⑤ 果 FRUITS	

① 三藥檳榔的樹幹。

Trunk of *Areca triandra* Roxb. ex Buch.-Ham.

② 莖簇生，纖細，直立至稍彎曲，葉痕明顯。

Stems clustered, slender, erect to slightly bent, with eminent leaf scars.

③ 羽狀複葉，羽片 17-20 對。羽片革質，暗綠色，具 2-6 條明顯葉脈，披針形，羽片下部小葉頂端漸尖，上部小葉頂端稍鈍形，具鋸齒，末端小葉與其他葉片合生。假葉柄着生於葉鞘基部。

Pinnately compound, leaflets 17-20 pairs. Leaflet blade leathery, dark green, evident veins 2-6, lanceolate, lower leaflets apex accumulate, upper leaflets apex slightly obtuse, toothed, terminal leaflets fused. Pesudopetioles sheathing.

④ 雌雄同株，雄花和雌花着生於同一花序上。穗狀花序，着生於葉鞘基部，多分枝。佛焰苞 1 朵，革質，早落。雄花較小，眾多，呈乳白色，雄蕊 3 條，雌花較大，簇生於花序基部。

Monoecious, male and female flowers on the same inflorescence. Spikes sprouting from the base of foliar sheaths, branched. Spathe 1, leathery, caducous. Male flowers small, numerous, creamy white, stamens 3, female flowers large, clustered at the base of inflorescence.

⑤ 果實呈橢圓形，成熟時轉為朱紅色。
Ellipsoid, scarlet-red when ripen.

生命力 VITALITY

　　三藥檳榔偏愛陽光充沛的環境和潮濕的土壤。與其他棕櫚科物種相較，三藥檳榔對低溫的耐受性高，即使氣溫降至 4℃，亦能茁壯成長。其對種植環境的要求低，故可減輕種植時對環境氣候控制的負擔。除了使用種子種植外，人們亦會從其根櫱（一種在主莖附近發芽的不定芽）為三藥檳榔進行無性繁殖。

Triandra Palm prefers sunny weather and moist soils. It is highlighted for the eminent tolerance to low temperature compared with other palm species, normally thriving even when the temperature is dropped to 4°C. The palm can be grown effortlessly due to its excellent acclimatising to versatile environments. Other than planting it from seeds, people tend to reproduce the palm vegetatively from its suckers, which are the adventitious shoots that sprout near to the main stem.

植物趣聞 ANECDOTE ON PLANTS

三藥檳榔與檳榔 Triandra Palm and *Areca catechu* (Betel Palm)：

　　三藥檳榔與檳榔皆屬於檳榔屬，而且具有相似的形態特徵。然而，它們的生長習性和果實外觀迥然不同，故不應把兩者誤認。檳榔屬於單一主莖棕櫚，最高可達 20 米，而三藥檳榔通常以灌木形式出現，其高度通常介乎 5 米左右。檳榔的成熟果實較大，呈黃橙色，而三藥檳榔的成熟果實則較小，呈朱紅色。

Two palms are the members of *Areca*, sharing outwardly similar traits. However, they should not be mistakenly diagnosed due to the distinct growing habits and fruit appearances. While Betel Palm is solitary palm that can grow to a towering 20 m, Triandra Palm is always shrubby and confined to 5 m. The mature fruits of Betel Palm are large and yellow orange, whereas those of Triandra Palm are smaller and scarlet.

備註 Remarks
本樹木學名根據世界植物線上網頁：
Scientific name of this tree is based on Plants of the World Online website:
https://powo.science.kew.org

霸王棕

Bismarck Palm | *Bismarckia nobilis* Hildebrandt & H.Wendl.

相片拍攝地點：遮打花園、香港迪士尼樂園
Tree Location: Chater Garden, Hong Kong Disneyland

名字由來 MEANINGS OF NAME

屬名 *Bismarckia* 是為了紀念 19 世紀第一任德國總理奧托・馮・俾斯麥（1815-1898）。種加詞 *nobilis* 即英文 noble 之意，代表貴族、崇高與偉大。值得留意的是，霸王棕是俾斯麥櫚屬的唯一品種。

The generic name *Bismarckia* was named after Otto von Bismarck (1815-1898), who was the first German Chancellor in the 19th century. The species name *nobilis* is pertinent to "noble" in English. Of note, Bismarck Palm is the only species in this genus.

本地分佈狀態 DISTRIBUTIONS	**外來物種** Exotic species
原產地 ORIGIN	馬達加斯加是原生霸王棕的唯一原產地。 The native Bismarck Palm is confined to Madagascar.
生長習性 GROWING HABIT	常綠，單一主莖棕櫚。高度可達 20 米。 Evergreen, solitary palm. Up to 20 m tall.

花果期 月份	1	2	3	4	5	6	7	8	9	10	11	12

花期：不詳。果期：不詳。
Flowering period: Unknown. Fruiting period: Unknown.

　　霸王棕擁有巨塔般的高度和筆直的樹幹，故被用作觀賞樹種種植在公園和花園中。擁有銀藍色葉片的霸王棕令人神搖目奪，故種植量較擁有綠色葉片的霸王棕多。這兩種霸王棕都很容易在香港迪士尼樂園尋見。

　　除觀賞價值外，霸王棕的樹莖和葉片是建造木板和屋頂的重要建築材料。而其莖內含澱粉，常被加工成西米食用。

Bismarck Palm is considered as a sought-after ornamental component for parks and gardens in view of its towering height and erect stem. The palm with silver blue fronds is always planted as a visual alternative of normal green palm for its more highlighted foliage colour. Bismarck Palm with two types of frond colours can be easily found In Hong Kong Disneyland.

Other than its gorgeous ornamental effect, the stem and leaves are qualified building materials for planks and roofs. The inner stem is starchy and often processed into sago.

辨認特徵 TRAITS FOR IDENTIFICATION

① 樹幹 TRUNK	② 樹皮 BARK	③ 葉 LEAVES
④ 花 FLOWERS	⑤ 果 FRUITS	

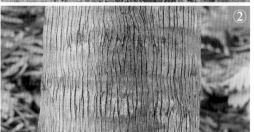

② 樹幹光滑，灰色，具不規則環狀葉痕及垂直裂隙。

Trunk grey, smooth, with irregular leaf scar rings and vertical fissures.

③ 葉片不對稱，呈淡銀藍色至藍綠色，掌狀，50-70 塊裂片，葉片向上拱起，頂端下垂。葉鞘腫脹，葉柄邊緣銳利或具小鋸齒，於葉柄頂端形成戟突，具厚重白色蠟質塗層。

① 霸王棕的樹幹。基部輕微隆起。幼時霸王棕樹幹基部具枯葉宿存。

Trunk of *Bismarckia nobilis* Hildebrandt & H. Wendl. Base slightly bulged. Trunk of the young palm always covered with dead persistent leaf bases.

Fronds asymmetrical flap, palmate, segments 50-70, pale silver blue to blue green, arching, drooping at the tip. Leaf sheath swollen, pseudopetiole sharp-edged or equipped with small teeth, with heavy white waxy coating. Hastula at the junction of frond and pseudopetiole.

④ 雌雄異株,單性花。圓錐花序,分枝成 2 條。雄性花序較長且較粗,花眾多而細小,呈乳白色。雌性花序較短。

Dioecious. Panicles, branched to 2 orders. Staminate inflorescences relatively longer and thicker, flowers many, small, creamy-white. Pistillate inflorescence shorter.

⑤ 核果長橢圓形至卵球形,成熟時轉為巧克力棕色。

Drupes oblong to ovoid, sepia at maturity.

生命力 VITALITY

　　霸王棕有不同的葉片顏色,分別為綠色和銀藍色。擁有銀藍色葉片的霸王棕喜歡全日照環境,在炎熱環境生長迅速。年幼時期的霸王棕無法抵禦寒冷天氣,而成熟後對低溫具耐受性。與擁有銀藍色葉片的霸王棕相比,擁有綠色葉片的霸王棕不太耐寒,更偏好熱帶氣候。兩種霸王棕皆偏好排水良好的土壤,以防止根部腐爛。

The fronds of Bismarck Palm are primarily dyed in green or silvery blue. The silvery blue Bismarck Palm loves full sunlight and grows particularly fast in hot weather; the mature palm shows moderate tolerance to low temperature. On the contrary, the green Bismarck Palm less acclimatise to coldness and prefers tropical climates. They both demand well-drained soils in order to prevent root rotting.

備註 Remarks

本樹木學名根據世界植物線上網頁:

Scientific name of this tree is based on Plants of the World Online website:

https://powo.science.kew.org

魚尾葵 又稱：青棕、假桄榔
Fishtail Palm | *Caryota maxima* Blume

相片拍攝地點：獅子會自然教育中心、屯門
Tree Location: Lions Nature Education Centre, Tuen Mun

名字由來 MEANINGS OF NAME

由於其葉形酷似魚尾，故被稱為魚尾葵。屬名 *Caryota* 源自希臘語 *caryon*，意為堅果。種加詞 *maxima* 有「最大」的含意，意指其樹身較其在親緣關係上接近，並被廣泛種植的短穗魚尾葵為大。

The generic name *Caryota* is derived from the Greek word *caryon*, meaning "a nut". The specific epithet *maxima* means "the largest", describing its relatively larger size when compared with Small Fishtail Palm (*Caryota mitis* Lour.), a closer species which is pervasively planted in Hong Kong. The leaves of the palm are akin to a fish tail, hence named as "Fishtail Palm".

本地分佈狀態 DISTRIBUTIONS	外來物種 Exotic species
原產地 ORIGIN	東南亞內陸、中國中南部及東南部。 Mainland Southeast Asia, and South-Central and Southeast China.
生長習性 GROWING HABIT	常綠，單一主莖棕櫚。高度可達 25 米。 Evergreen, solitary palm. Up to 25 m tall.

1	2	3	4	5	6	7	8	9	10	11	12

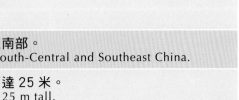

花果期月份

花期：本港五月至七月。果期：本港八月至十一月。
Flowering period: May to July in Hong Kong. Fruiting period: August to November in Hong Kong.

　　魚尾葵因具多種用途而備受青睞。其樹幹外圍堅韌，可用於製作拐杖及筷子。其莖部中心會產生大量澱粉，並富含蛋白質、維他命和纖維等對維持人體健康不可或缺的元素。其根部和葉片亦可採收，曬乾後可製成常見傳統中藥魚尾葵根和魚尾葵葉。魚尾葵根可治療筋骨痿軟、肝腎虧虛；魚尾葵葉可減少咳血、經血過多及便血等情況。其果實形似葡萄，但因其具毒性可引致皮膚過敏，故不可食用。

　　魚尾葵通常會替代短穗魚尾葵種植在開揚廣闊的空間。除了擁有粗壯的樹幹，魚尾葵開花時，一串串淡橙色淡雅的花朵隨風輕盈地搖曳，婀娜多姿。

Fishtail Palm has been appreciated for multiple uses. The periphery of the trunk is stiff and can be used for making walking sticks and chopsticks. Starch is fairly prolific in the inner stem, with abundant proteins, vitamins and fibres that are indispensable to maintain human health. The roots and leaves are harvested for and dried into the common traditional Chinese medicines, *Yuweikuigen* and *Yuweikuiye*. *Yuweikuigen* can treat flaccidity of extremities and liver-kidney deficiency while *Yuweikuiye* is useful to attenuate blood coughing, metrorrhagia and hemafecia. The fruits are grape-like, however, are not edible in respect of their toxicity that can induce skin allergy.

Fishtail Palm is always planted as a visual alternative of Small Fishtail Palm for wide open space. Apart from its robust trunk, when blossoms, it is blanketed in strings of pale orange flowers, which hang beautifully on the palm and swag lithely with the wind.

辨認特徵 TRAITS FOR IDENTIFICATION

① 樹幹 TRUNK	② 葉鞘 PETIOLES	③ 葉 LEAVES
④ 花 FLOWERS	⑤ 果 FRUITS	

① 魚尾葵的樹幹。樹幹灰色，單一主莖，無根蘗，具明顯環狀葉痕。

Trunk of *Caryota maxima* Blume. Trunk grey, without suckers, covered with eminent rings of leaf scars.

② 葉鞘具眾多黑色網狀纖維並簇生於葉基。

Petioles sheathing at the stem, with abundant reticulated black fibres clustering at the base.

③ 二回羽狀複葉，羽片 27 對，下垂，小羽片 12-27 對，互生至近對生。小羽片魚尾狀，暗綠色，末端小羽片較大，楔形，頂端具 2-3 裂，側生小羽片較小，菱形，外側筆直，內緣具不規則齒狀裂。

Bipinnately compound, pinnae 27 pairs, pendulous, pinnules 12-27 pairs, alternate to subopposite. Blade fish-tailed, dark green, terminal pinnules large, cuneate, apex 2 to 3-lobed, lateral pinnules small, rhombic, outer margin straight, inner margin irregular toothed.

④ 雌雄同株，雄花和雌花着生於同一花序。穗狀花序，腋生，多分枝，下垂，花序可長達 350 厘米。花朵呈淡橙色，雄花橢圓形，雄蕊眾多，雌花球狀。

Monoecious, male and female flowers on the same inflorescence. Spikes axillary, branched, pendulous, up to 350 cm long. Flowers pale orange, male flowers elliptic, stamens many, female flowers globose.

⑤ 果實呈球狀，成熟時由紅色轉為紫黑色。

Drupes globose, turning red to purplish black when ripe.

生命力 VITALITY

魚尾葵偏愛全日照和濕潤的土壤，並對強風具高耐受性。

Fishtail Palm has a preference for full sun and moist soil, and high wind tolerance.

植物趣聞 ANECDOTE ON PLANTS

魚尾葵與短穗魚尾葵 Fishtail Palm and *Caryota mitis* (Small Fishtail Palm)：

　　魚尾葵與短穗魚尾葵的葉形頗為獨特，故皆被認為是觀賞性高的棕櫚樹。短穗魚尾葵與魚尾葵的主要區別在於其顯而易見的樹身大小，短穗魚尾葵的樹身較魚尾葵小。除此之外，生長習性和果實外觀是防止誤認兩者的關鍵。與單一主莖、莖部堅硬的魚尾葵相比，短穗魚尾葵的莖部叢生，更為細長。魚尾葵的成熟果實呈紫黑色，短穗魚尾葵的成熟果實則呈紅色或橙色。

　　Two palms are regarded as graceful ornamental palms with their rather unique leaf shapes. Intuitively, Small Fishtail Palm differs from Fishtail Palm mainly due to its overall smaller size. However, the growing habit and fruit appearance are more likely the pivotal keys that thwart the mistaken diagnosis. Contrast to Fishtail Palm which the stem is always stiff and solitary, Small Fishtail Palm is prone to be slender and clustered. Whereas the mature fruits of Fishtail Palm are purplish black, those of Small Fishtail Palm are red or orange.

三角椰子

Triangle Palm | *Dypsis decaryi* (Jum.) Beentje & J.Dransf.

相片拍攝地點：青衣公園、香港公園
Tree Location: Tsing Yi Park, Hong Kong Park

名字由來 MEANINGS OF NAME

　　屬名 *Dypsis* 形容了其下垂的葉尖。種加詞 *decaryi* 是為了紀念此品種的第一位標本收集者——雷蒙・德卡里（1891-1973）而命名，他是一位把畢生都奉獻給馬達加斯加的法國科學家。因為其特殊的三列結構，故被俗稱為「三角椰子」。

　　The generic name *Dypsis* in Greek (*dyptein*) refers to its fronds with drooping tips. The specific epithet *decaryi* is in honour of the first specimen collector of the palm, Raymond Decary (1891-1973), a French botanist who dedicated his entire career in Madagascar. The palm is commonly known as "Triangle Palm" in respect of its leaves arranged in particular tristichous structure.

本地分佈狀態 DISTRIBUTIONS	外來物種 Exotic species
原產地 ORIGIN	馬達加斯加。 Madagascar.
生長習性 GROWING HABIT	常綠，單一主莖棕櫚。高度可達 10 米。 Evergreen, solitary palm. Up to 10 m tall.

花果期 月份	1	2	3	4	5	6	7	8	9	10	11	12

花期：不詳。果期：不詳。
Flowering period: Unknown. Fruiting period: Unknown.

　　許多香港的公園都設有棕櫚園。三角椰子是一種很容易在公園找到的棕櫚樹種。當下次逛公園時，不妨尋找一下其蹤影！

Many parks in Hong Kong have been designed with a Palm Garden. Triangle Palm is one of the predominant palms that can be spotted easily with its outstanding tristichous structure.

辨認特徵 TRAITS FOR IDENTIFICATION

① 樹幹 TRUNK	② 樹皮 BARK	③ 葉 LEAVES
④ 花 FLOWERS	⑤ 果 FRUITS	

① 三角椰子的樹幹。

Trunk of *Dypsis decaryi* (Jum.) Beentje & J.Dransf.

② 單一樹幹，圓狀，呈棕灰色，葉痕環狀，緊密。

Trunk solitary, round, brownish grey, prominently ringed with leaf scars.

③ 葉片排成 3 列，葉柄呈褐色，被棕黑色絨毛，呈三角狀。羽狀複葉，幾乎向上拱起，頂端下垂，呈綠色至銀綠色，小葉狹形，V 字型排列在葉軸上。

Leaves arranged in 3 rows, petiole brown, brown-black tomentose, triangular, arching almost upright, apex pendulous, leaflets bluish green to silvery green, narrow, v-shaped arrangement on the axis.

④ 雌雄同株異花，單性花。總狀花序緩緩分枝，呈黃色至綠色。

Monoecious. Spikes moderately branched, yellow to green.

⑤ 卵球形，成熟時由綠色轉為紅色。

Ovoid, turning green to red when ripe.

生態 ECOLOGY

儘管三角椰子在世界各地作為觀賞樹種被廣泛種植，但野生三角椰子集中分佈在馬達加斯加。在當地，其葉片用來製作屋頂，果實能生食，亦能加工成發酵飲品，種子則被收集到苗圃出售。除了以上因素，其棲息地亦受到嚴重破壞，如為了種植煙草而焚燒該地，導致野生三角椰子數量急劇下降。可惜的是，根據國際自然保護聯盟 2022 年的數據，野生種群中，成熟的三角椰子數量僅剩餘 999 棵，被評為「易危」。三角椰子是馬達加斯加當地動物的重要食物來源和庇護場所，其數量減少有可能危害當地的生物多樣性。

Although Triangle Palm is cultivated globally for ornamental purposes, the wild population is confined to only Madagascar. In its native range, its leaves are harvested for building roofs; the fruits can be eaten fresh and are often processed into fermented drinks; the seeds are collected by nurseries for selling. Coupled with excessive destruction to its habitats (e.g. burning the area for growing tobacco), the number of wild Triangle Palm has declined sharply. According to the data from the International Union for Conservation of Nature (IUCN) in 2022, regretfully, the wild population of mature Triangle Palm has restricted to only 999 and is rated as Vulnerable for the urgency of conservation. Triangle Palm serves as a primary food source and shelter for a myriad of local animals in Madagascar. The reduction of population is beyond all dispute deleterious to the local biodiversity.

生命力 VITALITY

三角椰子對乾旱、炎熱和半陰環境具高耐受性，但更偏愛全日照、濕潤的環境及沙質、排水良好的土壤。雖然成熟的三角椰子能抗火，但幼年時期對這種環境卻相當脆弱。

Triangle Palm is resistant to drought, heat and semi-shade, but good planting still prefers full sun with moist, sandy and well drained soils. Whereas mature Triangle Palm shows excellent tolerance to fire, the juvenile is rather fragile to such environments.

備註 Remarks

本樹木學名根據世界植物線上網頁：

Scientific name of this tree is based on Plants of the World Online website:

https://powo.science.kew.org

油棕

African Oil Palm, Oil Palm | *Elaeis guineensis* Jacq.

相片拍攝地點：青衣公園、逸東邨
Tree Location: Tsing Yi Park, Yat Tung Estate

名字由來 MEANINGS OF NAME

屬名 *Elaeis* 為希臘語，意指「油」。種加詞 *guineensis* 意指其鄰近幾內亞灣的原產地。

The generic name *Elaeis* comes from the Greek word, meaning "oil". The specific epithet *guineensis* refers to its origin near the Gulf of Guinea.

本地分佈狀態 DISTRIBUTIONS	**外來物種** Exotic species
原產地 ORIGIN	原生於剛果、幾內亞、奈及利亞及烏干達等非洲國家。 Native to the countries of Africa, such as Congo, Guinea, Nigeria and Uganda.
生長習性 GROWING HABIT	常綠，單一主莖棕櫚。高度可達 10 米。 Evergreen, solitary palm. Up to 10 m tall.

1	2	3	4	5	6	7	8	9	10	11	12

花果期
月份

花期：不詳。果期：不詳。
Flowering period: Unknown. Fruiting period: Unknown.

樹如其名，細小的油棕果實內蘊含極其豐富的油脂。油棕油分為兩種，取決於榨取的部分。榨取至中果皮的油脂為「棕櫚油」。與大豆等其他作物相比，油棕能提供更高產量（約 5.5 噸每公頃）、較高經濟價值以及更耐用的油。棕櫚油主要用作生物柴油和生物燃料，比起一般化石燃料更環保。至於由油棕種子提煉而成的則為「棕櫚仁油」，可加工成人造牛油等奶類製品，也可作為糖果、餅乾的原料。棕櫚油和棕櫚仁油皆可食用，亦是非洲當地人民的主要食材，為他們提供充足的營養，防止乾燥脫皮，擺脫皮膚病的困擾。

印尼和馬來西亞的棕櫚油產量佔全球一半，年收益超過十億美元。在如此可觀的棕櫚油產量和全球需求下，油棕產業發展失序膨脹。油棕適合種植於溫暖、陽光充沛和雨量充足的熱帶氣候地區。這些地區通常具豐富的生物多樣性，但為了開墾油棕種植園地，大規模伐木正以超乎想像的速度進行。例如 2007 年至 2013 年期間，秘魯棕櫚油業務急劇擴展，伐林數量增加了 11%。雖然秘魯的棕櫚油產業相對較小，但約六成用地集中在有大量生態區及高物種多樣性的亞馬遜流域附近。

濫伐林木是一場植物大屠殺，亦是無法逆轉的人為災難。森林資源是大自然賜予我們的珍貴財富，可惜人類並不愛惜這份本應用之不竭的的自然寶藏。

Like its common names, the fruits of African Oil Palm are profusely oily. From where the oil is extracted, it can be generally categorized into two types. "Palm oil" describes the oil yielded from the fruit mesocarp. Compared with other oil crops such as soybeans and corns, African Oil Palm is far and away the most prolific (approx. 5.5 tha^{-1}), producing economical and durable oils that are predominantly used as biodiesel and biofuels, which are suggested to be more environmentally friendly than ordinary fossil fuels. While the oil is extracted directly from the seeds is known as "palm kernel oil". It can be processed into dairy products like margarine and an ingredient of candies and biscuits. The palm oil and palm kernel oil are esculent and serve as a staple food of the local Africans, supporting daily energy and relieving skin diseases.

Indonesia and Malaysia are currently the most productive countries in the world, ruling more than half of the global palm oil production, with a profitable annual return of more than a billion USD. To attain the stunning demands of oils, the business of farming African Oil Palm has rampantly expanded and confined mainly to the tropical regions, where show warmer weather, abundant sunlight and copious precipitation. Since these regions are always highlighted as biodiversity hotspots, converting them into a homogenous land is beyond all dispute disastrous to the global biodiversity. For example, deforestation was increased 11% from 2007 to 2013 in Peruvian Amazonia due to the drastic expansion of the palm oil business in Peru.

Deforestation is fundamentally imperceptible to a massacre and is considered as catastrophic and irreversible. Forests are a huge wealth from our mother nature, but regrettably, humans are reluctant to receive it.

備註 Remarks
本樹木學名根據世界植物線上網頁：
Scientific name of this tree is based on Plants of the World Online website:
https://powo.science.kew.org

| ① 樹幹 TRUNK | ② 樹皮 BARK | ③ 葉 LEAVES |
| ④ 花 FLOWERS | | |

① 油棕的樹幹。

Trunk of *Elaeis guineensis* Jacq.

② 樹幹灰色，圓柱狀，樹幹基部具枯葉宿存。

Stem dark grey, cylindrical, with persistent old leaf bases.

③ 葉片掌狀分裂，螺旋排列。頂端下垂，葉片裂成 100 至 150 對，基部小葉退化成尖刺狀。葉柄被黃色絨毛，輕揉後脫落。

Pinnately compound, spirally arranged, leaflets 100-150 pairs, induplicate. Blade linear, tips drooping, entire, leaflets at the base modified and spiny. Pseudopetioles covered with yellow tomentose, shedding, margin teethed.

④ 雌雄同株，單性花。花序腋生，雄性花序為指狀穗狀花序，花朵眾多。雌性花序則為穗狀花序，近球形，柱頭三裂。

Monecious. Inflorescence axillary, male inflorescence spikelet, finger-like, many, female inflorescence spikelet, sub-globose, stigma 3-lobed.

註：本樹另有核果，卵球狀長橢圓形。

Remarks: Drupes, ovoid-oblong.

棍棒椰子

Spindle Palm, Palmiste Marron | *Hyophorbe verschaffeltii* (W.Bull ex J.Dix) H.Wendl.

相片拍攝地點：青衣公園
Tree Location: Tsing Yi Park

名字由來 MEANINGS OF NAME

屬名 *Hyophorbe* 是由 *hyo* 和 *phorb* 組成的希臘語混合字，分別意指「豬」和「飼料」，暗指棍棒椰子可用作飼料餵豬。為了紀念比利時園藝家 —— 安布魯瓦茲·維沙菲爾特（1825-1886），他生前痴迷於比利時的棕櫚苗圃，故取其種加詞為 *verschaffeltii*。因其樹幹粗壯如紡錘，故又俗稱為 Spindle Palm。

The generic name *Hyophorbe* is a blend of *hyo* and *phorb*, the Greek words that means "pig" and "feed", alluding to which the palm can be used for feeding pigs. The specific epithet *verschaffeltii* commemorates Ambroise Verschaffelt (1825-1886), a Belgian horticulturist who was obsessed with the nursery of palms in Belgium. Since the trunk is robust and akin to a spindle, it is also named as "Spindle Palm".

本地分佈狀態 DISTRIBUTIONS	外來物種 Exotic species
原產地 ORIGIN	棍棒椰子是印度洋上一個偏遠小島——羅德里格斯島的特有種。 Endemic to Rodrigues, a remote little island in the Indian Ocean.
生長習性 GROWING HABIT	常綠，單一主莖棕櫚。高度可達 9 米。 Evergreen tree, solitary palm. Up to 9 m tall.

花果期 月份	1	2	3	4	5	6	7	8	9	10	11	12

花期：不詳。果期：不詳。
Flowering period: Unknown. Fruiting period: Unknown.

① 樹幹 TRUNK	② 樹皮 BARK	③ 葉 LEAVES
④ 花 FLOWERS	⑤ 果 FRUITS	

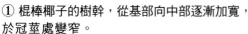

① 棍棒椰子的樹幹，從基部向中部逐漸加寬，於冠莖處變窄。

Trunk of *Hyophorbe verschaffeltii* (W.Bull ex J.Dix) H.Wendl., gradually widened from the base to the middle and narrowed near the crownshaft.

② 樹幹灰色，光滑，具明顯密集葉痕。

Trunk grey, smooth, with dense notable leaf scars.

③ 羽狀複葉，簇生於莖部頂端，小葉 50-70 對，互生。

Pinnately compound, clustered at the apex of stem, leaflets 50 to 70 pairs, alternate.

④ 雌雄同株，雄花和雌花簇生在不同花序上。大型穗狀花序，多分枝，從冠鞘基部長出。花朵芳香，呈乳黃色至橙紅色。

Monoecious, male and female flowers clustered on different inflorescences. Spikes large, branched, sprouting from the base of crownsheath. Flowers fragrant, creamy yellow to orange-red.

⑤ 漿果，圓柱形至長橢圓形，成熟時由紫棕色轉為黑色。

Berries cylindrical to oblong, purple-brown to black when ripe.

棍棒椰子與酒瓶椰子 Spindle Palm and *Hyophorbe lagenicaulis* (Bottle Palm)：

　　儘管棍棒椰子與酒瓶椰子具有相似的特徵，但可透過它們的樹身大小和樹幹外觀明確無誤地區分兩者。棍棒椰子較高，最高可達 9 米，而酒瓶椰子一般約高 3 米。此外，棍棒椰子的樹幹具明顯環狀葉痕，而酒瓶椰子的樹幹則具環狀葉痕及縱裂。

　　In spite of their similar traits, the two palms can be unmistakably diagnosed by their sizes and trunk appearances. Spindle Palm is relatively taller and its height can attain 9 m, while the height of Bottle Palm is generally limited to 3 m. Moreover, the trunk of Spindle Palm is only covered with eminent rings of leaf scars, while the trunk of Bottle Palm is covered with both leaf scar rings and longitudinal cracks.

酒瓶椰子屬是個單系屬，囊括 5 個物種，而且都是馬斯克林群島（包括羅德里格斯島、毛里求斯、留尼旺和眾多的火山遺跡）的特有種。棍棒椰子因其出色外觀而在全球範圍被廣泛培植。儘管酒瓶椰子屬的數量在世界各地急劇增加，它們的野生種群數量卻受到嚴重威脅，例如野生棍棒椰子目前僅存於羅德里格斯島。野生棍棒椰子於過去 150 年間發生了翻天覆地巨變，早期描述指出野生棍棒椰子遍佈羅德里格斯島（Balfour, 1879），但現時野生棍棒椰子數量減少至剩餘 19 株（IUCN, 2021）。目前，棍棒椰子在國際自然保護聯盟瀕危物種紅色名錄中被評為極危物種。這些年來該島嶼和島上的棍棒椰子究竟經歷了甚麼災難性事件，以致野生棍棒椰子走向成為極危物種的末路？

時光倒流回到幾個世紀前，16 世紀時阿拉伯和葡萄牙的水手首次登陸馬斯克林群島，很快便有不同開墾者接踵而至居住。開墾者登陸島嶼後，焚燒森林和砍伐木材以滿足資源需求，並引入野化動物，如山羊和兔子。這些不速之客踐踏了島上的土地，並啃食了包括棍棒椰子葉在內的植物，無可避免地將酒瓶椰子屬推向了滅絕邊緣。這如響鐘敲響了眾人對保育的覺悟。

20 世紀末，島上的山羊和兔子終被滅絕。這種短期措施有效復育了部分物種，例如酒瓶椰子，但野生棍棒椰子的數量仍在下降。保育生物學家目前正鍥而不捨地在羅德里格斯島種植大量棍棒椰子以恢復其種群數量。我們冀望他們奮鬥的努力能夠將這悲慘現象轉化為成功保育的故事，流傳後世。

Hyophorbe is a monophyletic genus of 5 species which are all endemic to the Mascarene Islands, including Rodrigues, Mauritius, Réunion and numerous volcanic remnants. The palms are cultivated globally for their excellent ornamental appearances. In spite of the dramatic increase of individuals in the world, their wild populations are severely threatened. For example, the wild Spindle Palm is currently confined to Rodrigues. It has suffered from a drastic change in the past 150 years. It was indicated in the early description that the individuals of Spindle Palm are all over the islands (Balfour, 1879), but the wild population is reduced to only 19 individuals at present (IUCN, 2021). Currently, Spindle Palm is rated as a Critically Endangered Species in the International Union for Conservation of Nature (IUCN) Red List of Threatened Species. You may wonder, what catastrophic events did the islands and palms experience?

Rolling back the clock to a few centuries before, the Mascarene Islands was first landed by the Arab and Portuguese sailors in the 16th century, and soon inhabited recurrently by different exploiters. The exploiters, after landing on the islands, burnt the forests and harvested woods to satisfy their needs of resources; they introduced feral animals e.g. goats and rabbits into the islands. The animal intruders trampled every hectare of the island and foraged greens, including leaves of Spindle Palm, and inevitably drove *Hyophorbe* to the edge of extinction. This sounded an alarm to conservation.

In the late 20th century, the goat and rabbit were exterminated; this short-term practice effectively recovers some of the species e.g. Bottle Palm (*H. lagenicaulis*), while the wild population of Spindle Palm still remains declining. Conservation biologists are currently endeavouring in regenerating the population by planting mass Spindle Palm in Rodrigues. We hope their hard work can turn this tragic phenomenon into a successful conservation story.

備註 Remarks
本樹木學名根據世界植物線上網頁：
Scientific name of this tree is based on Plants of the World Online website:
https://powo.science.kew.org

加拿利海棗 又稱：加那利刺葵

Canary Island Date Palm, Pineapple Palm | *Phoenix canariensis*
Chabaud

相片拍攝地點：香港迪士尼樂園
Tree Location: Hong Kong Disneyland

名字由來 MEANINGS OF NAME

種加詞 *canariensis* 和中文名稱「加拿利海棗」意指其原產地為加拿利群島。

The specific epithet *canariensis* and the common name Canary Island Date Palm mean that their place of origin is the Canary Islands.

應用 APPLICATION

加拿利海棗是一種廣為人知的觀賞性培育品種。香港迪士尼樂園把加拿利海棗種植在神奇道，營造出壯觀的景象。雖然加拿利海棗具觀賞價值，但因造價昂貴，故對公園綠化來說十分奢侈。這棵樹不僅價格高昂，每棵身價超過 10 萬港元，而且對濕度十分敏感，根據以往報告，有部分加拿利海棗在香港因受病原體感染和高濕度影響而變得衰弱，可見此品種在香港難以維護。

本地分佈狀態 DISTRIBUTIONS	外來物種 Exotic species
原產地 ORIGIN	加拿利群島（由 7 個座落於摩洛哥西南部的島嶼組成）。 The Canary Islands, where are an archipelago composed of seven islands in the Southwest Morocco.
生長習性 GROWING HABIT	常綠，喬木狀棕櫚。高度可達 20 米。 Evergreen, tree-like palm. Up to 20 m tall.

1	2	3	4	5	6	7	8	9	10	11	12
								9			

花果期月份

花期：本港九月。果期：不詳。
Flowering period: September in Hong Kong. Fruiting period: Unknown.

Canary Island Date Palm is a notable ornamental palm cultivated worldwide. In Hong Kong, the palms are aligned in rows on the Magic Road of the Hong Kong Disneyland to create a staggering scene. Although the ornamental value of the tree is high, it is an extravagance to public greening spaces. Not only dose the tree cost for more than HK$ 100,000, but it is also susceptible to high moisture content. According to the previous reports, some of the palms became feeble due to the pathogenic infection and frequent rainfall in Hong Kong. It is difficult to maintain the species in Hong Kong.

① 樹幹 TRUNK	② 樹皮 BARK	③ 葉 LEAVES
④ 花 FLOWERS	⑤ 果 FRUITS	

① 加拿利海棗的樹幹。

Trunk of *Phoenix canariensis* Chabaud.

② 樹幹直立，具明顯及巨大鑽石狀環狀葉痕。

Stem erect, leaf scar obvious and large, arranged like diamonds.

③ 羽狀複葉。小葉大約200對，條狀。葉緣全緣，從同一平面延伸。基部小葉具尖刺。葉面光滑，老時從綠色轉為黃色。

Pinnately compound, leaflets about 200 pairs, linear, entire, spreading in the same plane, leaflets at base modified into sharp spines, glabrous, turning from green to yellow when old.

④ 雌雄異株。穗狀花序腋生，多分枝。花乳黃色。雄花序直立，雌花序下垂。

Dioecious. Spikes axillary, multi-branched. Flowers milky yellow. Male inflorescences erect, female inflorescences pendulous.

⑤ 果實橢圓形，成熟時轉為紅橙色，種子 1 粒。

Fruits ellipsoid, turning reddish orange when mature, 1-seeded.

生態 ECOLOGY

由於在加拿利海棗附近並沒有蜜蜂和甲蟲等有效的傳粉媒介的發現記錄，人們相信野生的加拿利海棗依靠風媒傳粉。此外，加拿利海棗沒有分枝，表示不能夠依靠折枝的方法單性繁殖，需要依靠種子繁衍後代。它的種子能吸引雀鳥，例如烏鶇會幫忙傳播種子。棗核甲蟲和樹皮甲蟲則可能是捕食者，妨礙種子產生。

The palm in the wild is believed to be wind-pollinated with the paucity of observation record of effective pollinators, such as bees and beetles, near the trees. Moreover, offshoot is absent in *P. canariensis*, and hence asexual reproduction is not possible, alluding to the necessity of producing offspring through seeds. Its seeds can attract birds. For example, *Turdus merula* (Blackbirds) help seed dispersal. Meanwhile, *Coccotrypes dactyliperda* (Date Stone Beetle) and *Dactylotrypes longicollis* (Bark Beetle) are possible predators that can suppress seed yield of the palm.

生命力 VITALITY

加拿利海棗對乾旱、風、高鹽度和貧瘠土壤具有良好的耐受性。但是它需要炎熱天晴的生長環境，加拿利海棗在寒冷或過度灌溉的情況下，生長會受阻礙。同時，它需要充足的生長空間。最高的加拿利海棗為 36 米，位於加拿利群島。

It shows excellent resilience to drought, wind, high salinity and barren soils. However, its growth requires hot and sunny environments, but is impeded under cold weather and overirrigation. Moreover, it needs a large growing space. The tallest Canary Island Date Palm is found in the Canary Islands, with a height of 36 m.

海棗
Date Palm | *Phoenix dactylifera* L.

相片拍攝地點：香港迪士尼樂園
Tree Location: Hong Kong Disneyland

種加詞 *dactylifera* 是由希臘詞 *dactylus* 和拉丁詞 *ferrous* 組成的混合字，意指「具手指」，暗指海棗能長出長橢圓形和手指狀的果實。

The specific epithet *dactylifera* is a blend of the Greek word *dactylus* and Latin word *ferous*, together known as "finger-bearing", alluding to its capacity of producing oblong and finger-like fruits.

本地分佈狀態 DISTRIBUTIONS	外來物種 Exotic species
原產地 ORIGIN	阿拉伯半島、伊朗、伊拉克和巴基斯坦。 Arabian Peninsula, Iran, Iraq and Pakistan.
生長習性 GROWING HABIT	常綠，單一主莖棕櫚。高度可達 5 米。 Evergreen, solitary palm. Up to 5 m tall.

花果期月份	1	2	3	4	5	6	7	8	9	10	11	12

花期：本港三月至四月。果期：本港九月至十月。
Flowering period: March to April in Hong Kong. Fruiting period: September to October in Hong Kong.

| ① 樹幹 TRUNK | ② 樹皮 BARK | ③ 葉 LEAVES |
| ④ 花 FLOWERS | ⑤ 果 FRUITS | |

① 海棗的樹幹。基部腫脹，單一主莖。

Trunk of *Phoenix dactylifera* L. Base swollen, stems solitary.

② 灰棕色，具菱形葉柄基部宿存。

Greyish brown, covered with persistent diamond-shaped leaf bases.

③ 羽狀複葉，可長達 4 米，葉軸每側約具 200 片小葉。小葉葉片劍形，呈灰綠色，互生或對生，在葉軸上排成數個平面。小葉基部為刺狀。

Pinnately compound, to 4 m long, around 200 leaflets per side of rachis. Leaflet blade sword-shaped, greyish green, alternate or opposite, arranged in several planes on the rachis. Leaflets at the base modified, spiny.

④ 雌雄異株。穗狀花序，腋生，多分枝，直立，雌性花序後期轉為下垂。花朵呈乳黃色。

Dioecious. Spikes axillary, branched, erect, female inflorescence later becoming pendulous. Flowers creamy-yellow.

⑤ 卵球形至長橢圓形，成熟時由橙色轉為紫黑色，種子 1 粒。

Drupes ovoid to oblong, turning from orange to purplish black when mature, 1-seeded.

種子休眠 Seed dormancy：

我們通常以種子發芽後開始估算一株植物的生命週期，若果把種子發芽前的時間也計算在內，那麼植物的生命週期有多長呢？答案或許會令你大吃一驚。

海棗自距離至今 6000 年至 6700 前，因其美味及無與倫比的藥用價值，而在美索不達米亞和波斯灣上部被栽培。發芽後，海棗的生命週期可長達 100 年或以上。近年，科學家發現世界最長壽的海棗位於以色列的馬薩達。發現這棵海棗時，它依然處於種子的狀態並散落在瓦礫下。令人震驚的是，根據放射性碳定年法的結果，這粒種子已經休眠了 2000 年！它在 2008 年成功發芽，目前種植於 Arava Institute（音譯為阿拉瓦環境研究機構），並受到 24 小時監護。因其出乎意料、極長的休眠期，故被稱為「瑪土撒拉」——一位享年 969 歲的聖經人物。

海棗是植物活力的象徵。在永無止境的進化和物競天擇下，植物裝備了多種「武器」，以適應瞬息萬變的環境。種子休眠可能是具有種子的植物最強大的機制，能夠靜候合適時機，待他日環境合適時才萌芽。當我們回望石炭紀時期，當時整個世界一部分是樹沼，孢子植物如蕨類植物為優勢種。種子植物已經如先知一般預測到環境的變化。當時種子植物首先扎根於沼澤高地，遠離競爭激烈的孢子植物。與需要水分繁衍後代的孢子植物不同，種子能忍耐並等待更好的時機發芽。踏入二疊紀後，氣候變得乾燥，種子植物的數量便激增並成為優勢種。

We usually begin to estimate a plant's life span when it germinates from a seed, but how long can be the life span of a plant if the time before seed germination are also taken into account? The answer can be surprising.

Date Palm has been cultivated since 6700 to 6000 B.P. in Mesopotamia and Upper Arabian Gulf, in view of its delicious taste and wonderful medicinal effects. After the germination, the palm can live for a general 100 years or above. In recent years, scientists found the world's most aged Date Palm in Masada, Israel. When the palm was found, it still remained as a seed and laid up under heaps of rubble. According to the radiocarbon dating result, astonishingly, the seed could have already been dormant for 2000 years! It germinated successfully in 2008 and is currently planted in the Arava Institute, under 24-hour monitoring. In respect of its unexpectedly long dormancy, it is also named as "Methuselah", a biblical character who died at age of 969.

Data Palm is emblematic of plants' vigour. Under the endless evolution and selective pressure, plants have been armed with multiple "weapons" so they can adapt to dynamic environments. Seed dormancy is probably one of the most powerful mechanisms equipped by seed-bearing plants. The seed waits for a suitable time to germinate. We can take a glimpse into the Carboniferous period, which parts of the world were still swampy and dominated by spore plants (e.g. ferns). Seed plants forecasted the change of environment. They first rooted in uplands of the swamps and stayed away from the drastic competition from spore plants. Different from spore plants which require water to produce the next generations, seeds can endure and sprout when they spot a better-off moment. When the earth stepped into the Permian, the period when the climate was getting drier, seed plants then proliferated and became dominant.

刺葵 又稱：台灣海棗

Spiny Date Palm, Mountain Date Palm, Formosan Date Palm | *Phoenix loureiroi* Kunth

相片拍攝地點：青衣公園
Tree Location: Tsing Yi Park

名字由來 MEANINGS OF NAME

為了紀念葡萄牙耶穌會傳教士暨植物學家——若昂·德·洛雷羅（1717-1791），故其種加詞為 *loureiroi*。

The specific epithet *loureiroi* is in honour of João de Loureiro (1717-1791), who was a Portuguese Jesuit missionary and botanist.

應用 APPLICATION

在塑膠製掃帚未普及前，人們通常用刺葵的老葉做掃帚，俗稱為「糠榔帚」。菲律賓人通常會將其幼葉曬乾並編織成雨衣。其果實可生食，並具有抗癌及抗氧化作用。

Before the prevalence of plastic brooms, people used its old leaves as brooms, known as "Kang-lang broom". In the Philippines, its juvenile leaves are dried and woven into raincoats. Its fruits can be eaten fresh and are anticancer and antioxidant.

本地分佈狀態 DISTRIBUTIONS	原生物種 Native species
原產地 ORIGIN	中國東南部省份及其他東南亞國家，如老撾、巴基斯坦、菲律賓等。 Southeast China and other Southeast Asian countries, e.g. Laos, Pakistan, the Philippines.
生長習性 GROWING HABIT	常綠，單一主莖或聚生棕櫚。高度可達 5 米。 Evergreen, solitary or clustered palm. Up to 5 m tall.

1	2	3	4	5	6	7	8	9	10	11	12	花果期月份

花期：本港四月至五月。果期：本港六月至十月。
Flowering period: April to May in Hong Kong. Fruiting period: June to October in Hong Kong.

① 樹幹 TRUNK	② 樹皮 BARK	③ 葉 LEAVES
④ 花 FLOWERS	⑤ 果 FRUITS	

① 刺葵的樹幹。

Trunk of *Phoenix loureiroi* Kunth.

② 莖呈暗褐色，節間極短，具菱形葉柄基部宿存。

Stem dark brown, internodes very short, with persistent diamond-shaped petiole base.

③ 羽狀葉片長達 2 米，每邊葉軸具 40-130 條羽片，線狀，單生或 2-3 片排列在不同平面上。羽片基部具刺。

Pinnately compound, to 2 m, leaflets 40-130 pairs, linear, solitary or 2-3 arranged in different planes, base modified, spiny.

④ 雌雄異株。花序直立，具分枝，最先出的葉呈黃綠色，花朵呈黃白色。雄花花瓣 3 片，雄蕊 6 條。雌花卵球形。

Dioecious. Inflorescences erect, branched, prophyll yellowish green, flowers yellowish white. Male flowers petals 3, stamens 6. Female flowers ovoid.

⑤ 果實長橢圓形，成熟時由橙色轉為黑色或紫黑色。

Drupes oblong, turning from orange to black or purplish black when ripe.

刺葵通常生長在海拔低於 1700 米的草原、林地和沿海地區。為了在大自然中掙扎求存，它已進化成對全日照、乾旱、澇漬、強風及鹽分具高耐受性。

其雄花深受甲蟲喜愛；果實是多種鳥類和哺乳類動物的部分食源。

Spiny Date Palm is prone to grow in grasslands, woodlands and coastal areas, usually below 1700 m above sea level. To acclimatize to the environments, it has evolved superior tolerances to full sunlight, drought, dampness, wind and moderate salinity.

Its male flowers are beloved by beetles. Its fruits serve as a delicacy for a wide range of birds and mammals.

香港的刺葵屬 *Phoenix* spp. in Hong Kong：

在香港，刺葵屬被視為觀賞棕櫚樹並廣泛種植。它們的名稱和形態都大同小異，故常被混淆。刺葵在部分地方被視為台灣海棗，但它目前被認為是刺葵的別名。值得一提的是，可能是因為錯誤種植或錯誤識別，在香港大多數標示為台灣海棗的棕櫚樹實際上是銀海棗。

要區分它們，可從樹木大小、樹皮及葉片入手。刺葵的樹身較矮小，銀海棗則堅硬且高聳。刺葵的樹皮呈暗褐色，密被葉柄基部宿存，銀海棗的樹皮則呈灰色，被散落的葉柄基部宿存。刺葵的葉片呈綠色，而銀海棗的葉片則呈藍綠色。

刺葵屬中只有刺葵是唯一原生並廣泛分佈於香港。與市區公園中看到的刺葵不同的是，野生刺葵通常以灌木形式出現，不一定有短莖。

Phoenix spp. are widely grown in Hong Kong for ornamental purpose but are often confused with their comparable names and morphologies. For example, Formosan Date Palm (*P. hanceana*) is sometimes considered as a distinct species of Spiny Date Palm, but the former is treated as a synonym of *P. loureiroi*. Notably, most of the palms in Hong Kong are mistakenly labelled as Formosan Date Palm and should be *P. sylvestris* (Wild Date Palm) instead.

We can use size, bark and leaf to distinguish the two palms, Spiny Date Palm is always shorter while Wild Date Palm is stiff and towering. The stem of Spiny Date Palm is dark brown and covered with dense persistent petiole bases, while the Wild Date Palm is grey and covered with scattered persistent petiole bases. The leaves of Spiny Date Palm are green, while those of Wild Date Palm are blue-green.

Spiny Date Palm is the only native Phoenix species in Hong Kong and is widely distributed. Unlike the one that you can see in urban parks, the wild Spiny Date Palm is always bush-like and may not with a short stem.

銀海棗 又稱：林刺葵

Wild Date Palm, Sugar Date Palm | *Phoenix sylvestris* (L.) Roxb.

相片拍攝地點：九龍公園、香港迪士尼樂園
Tree Location: Kowloon Park, Hong Kong Disneyland

名字由來 MEANINGS OF NAME

屬名 *Phoenix* 來自古希臘語，用於描述海棗。種加詞 *sylvestris* 意指其棲息地為森林。俗名 Wild Date Palm（野生海棗樹）是因為它外形與海棗（*Phoenix dactylifera*）如出一轍，而 Sugar Date Palm 則描述其果實糖分極高。

The generic name *Phoenix* comes from the ancient Greek word for describing a date palm. The specific epithet *sylvestris* means "growing in a forest". The common name "Wild Date Palm" is given with its comparable appearance to Date Palm (*P. dactylifera*) while "Sugar Date Palm" is to describe its high sugar content.

本地分佈狀態 DISTRIBUTIONS	**外來物種** Exotic species
原產地 ORIGIN	僅分佈於印度及其周邊地區，如緬甸、尼泊爾和巴基斯坦。 India and the neighbouring countries, such as Myanmar, Nepal and Pakistan.
生長習性 GROWING HABIT	常綠，單一主莖棕櫚。高度可達 16 米。葉片密集成半球形樹冠，約有 100 片羽狀葉。 Evergreen, solitary palm. Up to 16 m tall. Crown hemispherical, very dense, with about 100 leaflets.

花果期 月份	1	2	3	4	5	6	7	8	9	10	11	12
									9	10		

花期：不詳。果期：本港九月至十月。
Flowering period: Unknown. Fruiting period: September to October in Hong Kong.

銀海棗具豐富食用及藥用價值。在印度和印尼，當地人鍾愛直接飲用銀海棗莖部中甜的汁液，或加工成粗糖或酒精飲品。汁液中含有大量鐵質和維他命（例如B12），具營養價值，並能用於治療貧血。其果實可食用，有時加工成果凍或果醬，亦可製作成補品或鎮痛劑以緩解痛楚。此外，葉片可緩解眼部發炎，而根部則能改善神經衰弱。除醫藥用途外，由於其木材可用於建造橋樑和碼頭，故當地人廣泛種植銀海棗，以滿足日常建設需求。

銀海棗對貧瘠土壤、乾旱和洪水具高度耐受性，故在管理上如運諸掌。加上銀海棗擁有茂密的樹冠和令人眼前一亮的樹形，是上佳的觀賞品種。銀海棗在香港被廣泛種植，同時是市民公認喜愛的刺葵屬品種。

Wild Date Palm shows high dietary and medicinal values. In India and Indonesia, the locals love to directly drink the sugary sap tapped from the stem of Wild Date Palm or process it into a jaggery or alcoholic beverage. The sugary sap contains rich Iron and vitamins (e.g. Vitamin B12) and with high nutritional value. It can also treat anemia. The fruits can be eaten fresh but are often processed into jellies and jams; they can also serve as a tonic or an analgesic to release pain. In addition, the leaves can relieve eye inflammation while the roots can ameliorate nervous debility. Apart from the medical usage, Wild Date Palm is planted locally for supporting the daily construction needs; the wood can be used for building bridges and piers.

Wild Date Palm shows high tolerance to barren and waterlogged soils. Coupled with the dense foliage and decent tree form, it is an excellent ornamental species. The palm is widely planted and considered as one of the beloved Phoenix spp. in Hong Kong.

① 樹幹 TRUNK	② 樹皮 BARK	③ 葉 LEAVES
④ 果 FRUITS		

① 銀海棗的樹幹。基部腫脹。

Trunk of *Phoenix sylvestris* (L.) Roxb. Base swollen.

② 莖呈暗褐色，節間極短，具菱形葉柄基部宿存。

Stem grey, with very short internodes, with persistent diamond shaped old leaf base.

③ 葉片羽狀分裂，小葉劍形，互生或對生，不規則簇生，經常 2-4 對排列在葉軸上。

Pinnately compound, leaflets sword-shaped, alternate or opposite, irregularly fascicled, always 2-4 pairs arranged in several planes on the rachis.

④ 核果倒卵形，成熟時由綠色轉為橙黃色。

Drupes obovoid, turning from green to orange yellow when mature.

註：本樹另有花，雄性花序挺立，約 20-30 厘米長，雌性花序先挺立，後下垂。

Remarks: Male inflorescences erect, around 20-30 cm long; Female inflorescences first erect, later drooping.

備註 Remarks

本樹木學名根據世界植物線上網頁：

Scientific name of this tree is based on Plants of the World Online website:

https://powo.science.kew.org

棕櫚

Windmill Palm | *Trachycarpus fortunei* (Hook.) H. Wendl.

相片拍攝地點：賽馬會德華公園、青衣公園
Tree Location: Jockey Club Tak Wah Park, Tsing Yi Park

名字由來 MEANINGS OF NAME

　　屬名 *Trachycarpus* 是由 *trachys* 和 *karpos* 組成的希臘語混合字，意為「粗糙果實」。為表揚著名蘇格蘭植物學家 —— 羅伯特・福鈞（1812-1880）將首個棕櫚標本送到邱園（英國皇家植物園），種加詞特此取名為 *fortunei*。葉片呈風車狀排列於棕櫚頂端，故又名 Windmill Palm。

The generic name *Trachycarpus* is a blend of Greek words *trachys* and *karpos*, collectively meaning "rough fruit". The specific epithet *fortunei* is named after Robert Fortune (1812-1880), an eminent Scottish botanist who sent the first specimen of this species to Kew Garden. The leaves are windmill-like arranged at the apex of stem, hence named as "Windmill Palm".

本地分佈狀態 DISTRIBUTIONS	外來物種 Exotic species
原產地 ORIGIN	中國中南部和緬甸。 South-Central China and Myanmar.
生長習性 GROWING HABIT	常綠，單一主莖棕櫚。高度可達 12 米。 Evergreen, solitary palm. Up to 12 m tall.

1	2	3	4	5	6	7	8	9	10	11	12	花果期月份

花期：本港五月至六月。果期：本港八月至九月。
Flowering period: May to June in Hong Kong. Fruiting period: August to September in Hong Kong.

棕櫚的莖部帶有凌亂美，故被廣泛種植於公園和花園。它可單一種植或與其他植物合併成一片別致的景觀。

除了觀賞用途外，棕櫚還有其他實際用途。其樹幹耐用且抗濕程度高，故可用於建造涼亭或其他建築物。其葉片基部的纖維堅韌且抗濕程度高，可用作繩子、籃子及漁網的原料。其葉柄下部經移除纖維及乾燥後可用作傳統中藥——「棕櫚」。棕櫚富含精油，有效減緩高血壓和中風風險。

Windmill Palm is broadly cultivated in parks and gardens in view of its handsome shaggy stem. It can be either planted in solitary or mixed with other greening components to draw an exquisite landscape.

The palm also serves multiple uses other than ornamental function. The trunk is durable and highly resistant to moisture, hence can be used for making pavilions and other constructions. The fibres clustered at the leaf bases are firm and resistant to moisture; they are raw materials for cordage, baskets and fishing nets. The lower part of petioles after removing fibres and drying can be applied as the traditional Chinese medicine, known as *Zonglü*. The palm contains abundant essential oils, which can attenuate hypertension and prevent stroke.

辨認特徵 TRAITS FOR IDENTIFICATION

① 樹幹 TRUNK	② 樹皮 BARK	③ 葉 LEAVES
④ 花 FLOWERS	⑤ 果 FRUITS	

① 棕櫚的樹幹。

Trunk of *Trachycarpus fortunei* (Hook.) H.Wendl.

② 樹幹暗褐色，被老葉柄基部宿存，纖維質。

Trunk dark brown, covered with old petiole bases, fibrous.

③ 葉片扇形，掌狀淺裂，深裂成 30-50 片，線狀、不重疊，頂端具 2 淺裂，下垂。葉柄具銳齒。

Blade fan-shaped, palmatifid, deeply 30-50 lobed, linear, induplicate, apex shallowly 2-lobed, drooping. Petioles sharply dentate.

④ 常雌雄異株。穗狀花序，堅硬，具 2 個或多分枝花序，每 2-4 朵簇生於花序，稀單生，花序在開花前包裹於佛焰苞內。花朵較小，雄花呈黃綠色，雌花呈淡綠色。

Usually dioecious. Spikes stiff, 2 or multi-branched, flowers 2-4 clustered, rarely solitary, inflorescence enclosed in multi-spathes before blooming. Flowers small, male flowers yellowish green, female flowers pale green.

⑤ 核果球形至長橢圓形，成熟時由黃綠色轉為藍色。

Drupes spherical to long elliptic, turning from yellow green to blue when ripe.

生命力 VITALITY

　　棕櫚偏愛營養豐富且潮濕的土壤。雖然棕櫚對水需求高，但它不能忍受澇漬土壤，水分過多會導致根部腐爛，故其偏愛排水良好的土壤。棕櫚於寒冷、強風、乾旱和空氣受到污染的環境中能茁壯成長，故備受青睞。

Windmill Palm prefers nutritious and moist soils. Despite its great demand for water, it still prefers well-drained soils and is susceptible to waterlogging that can cause root rotting. Windmill Palm is appreciated for its resilience to cold, windy, drought and air-polluted environments.

大絲葵 又稱：華盛頓葵

Petticoat Palm, Mexican Fan Palm | *Washingtonia robusta* H. Wendl.

相片拍攝地點：青衣公園、荔枝角公園
Tree Location: Tsing Yi Park, Lai Chi Kok Park

名字由來 MEANINGS OF NAME

屬名 *Washingtonia* 是紀念美國第一任總統 —— 喬治‧華盛頓（1732-1799）。種加詞 *robusta* 形容其粗壯的莖和對環境具高適應力。

The generic name *Washingtonia* is in honour of George Washington (1732-1799), who was the first president of the United States. The specific epithet *robusta* depicts its robust stem and great adaptability to multiple environments.

應用 APPLICATION

大絲葵的葉片經常應用在建造屋頂、製作籃子或涼鞋；葉片纖維常用於製作繩索。由於大絲葵擁有堅固的結構，故是一種被廣泛種植的街道觀賞樹。

The leaves are used for building roofs, making baskets and sandals. The fibre is used for making cord. Petticoat Palm is also a wonderful roadside and ornamental palm due to its robust structure.

本地分佈狀態 DISTRIBUTIONS	外來物種 Exotic species
原產地 ORIGIN	野生僅分佈於墨西哥西北部。 The wild Petticoat Palm is confined to Northwestern Mexico.
生長習性 GROWING HABIT	常綠，單一主莖棕櫚。高度可達 27 米。 Evergreen tree, solitary palm. Up to 27 m tall.

花果期 月份	1	2	3	4	5	6	7	8	9	10	11	12

花期：不詳。果期：不詳。
Flowering period: Unknown. Fruiting period: Unknown.

① 樹幹 TRUNK	② 樹皮 BARK	③ 葉 LEAVES
④ 花 FLOWERS	⑤ 果 FRUITS	

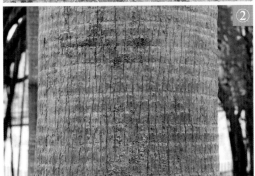

① 大絲葵的樹幹。基部腫脹。

Trunk of *Washingtonia robusta* H.Wendl. Base swollen.

② 樹幹棕灰色，常具下垂枯葉十字交錯於樹幹上，葉痕明顯，縱裂紋不明顯。

Stem brownish grey, drooping dead leaves remained at the upper stem sometimes, leaf scars conspicuous, with inconspicuous longitudinal cracks.

③ 葉片掌狀分裂，於葉片 3 分之 2 位置開裂，葉背基部具脫落性絨毛，頂端下垂，葉柄粗壯，具紅棕色刺。

Fronds palmately divided at 2/3 of the frond, abaxially deciduously tomentose, tips drooping. Pesudopetioles stout, armed with reddish brown spines.

④ 雌雄同株，雙性花。圓錐花序，具 5-6 條長分枝，下垂，呈奶油色。

Monecious, bisexual flowers. Spikes axillary, 5-6 branched, large, pendulous, flowers cream.

⑤ 果實球形至橢圓形，成熟時轉為棕黑色。

Drupes globular to elliptic, turning brown-black at maturity.

　　雙子葉植物次級生長時會在樹幹水平增長，闊度增加，使樹幹最外層覆蓋着堅硬的樹皮組織。不同的是，單子葉植物只進行初生生長。換言之，它們不能用樹皮保護莖部。那它們是否極之脆弱？棕櫚樹雖為單子葉植物，但總能抵禦大風蹂躪。沒有樹皮的它們是如何在這般惡劣的環境中掙扎求存？

　　大絲葵在一些研究中為模範棕櫚品種，有助揭開背後機制。科學家們指出，維管束帽是防止樹木倒塌的關鍵。在解釋維管束帽前，我們應先了解單子葉植物中的維管束系統。植物透過維管束運輸養分和水分至全棵植物。一般而言，維管束是一束初生組織，主要由木質部（水分運輸）和韌皮部（養分運輸）組成。它們數量眾多，在雙子葉和單子葉植物中的分佈各有不同。將莖部水平剖開，能看到雙子葉植物的維管束統一排成環狀，在單子葉中則分散排序。棕櫚中的維管束通常呈分散分佈，但仍能觀察其放射性結構，有較多維管束集中在莖部外圍，較少集中在莖部中央。

　　回到之前的問題，甚麼是維管束帽？簡而言之，它是一層厚而堅硬的細胞，通常集中在維管束邊緣，能支撐植物穩立。科學家們觀察到在大絲葵莖部邊緣具有不同維管束，當中有不少特定維管束帽設計。維管束的最外層由完全木質化的維管束帽層組成，能支撐莖部，作出適當的彎度，以抵禦強風。當細胞移動到莖部中心時，維管束帽的剛度便會降低。這種演化被認為會增加全株植物的擺動並消耗風的能量。這些機制避免了大絲葵被強風吹毀，令其在強風中屹立不倒。

Secondary growth of dicots confers girth to stems and allows the stem to be covered by hard bark tissues towards its periphery. Unlike dicots, monocots are limited to only primary growth. In other words, their stems are not protected by bark. Then, are they very fragile? Palms as monocots, however, can resist strong winds. Without the protection of bark, how can they survive in such harsh environments?

Petticoat Palm, as a model palm species in some studies, helps unravel the mechanisms behind. The scientists indicated that the bundle cap layer is important to prevent the palm from falling. Before explaining what a bundle cap is, we should understand the vascular bundle system in monocots. Plants transport nutrients and water throughout their bodies through vascular bundles. In general, a vascular bundle is a bunch of primary tissues consisting mainly of the xylem (water transport) and phloem (nutrient transport). Vascular bundles are abundant and distributed in different patterns between dicots and monocots. By transverse cut through the stems, we can see that the vascular bundles in dicots are arranged uniformly in a ring shape, while they are dispersed in monocots. Usually in palms, the vascular bundles are scattered but a radial structure can also be observed. They are accumulated towards the periphery of the stem.

Back to our previous question, what is a bundle cap? Concisely, it is a layer of thick and rigid cells, which is always located at the edge of vascular bundles. This layer is functionally important to support a plant for standing stiffly. In Petticoat Palm, scientists observed that the vascular bundles near the periphery of stem are equipped with completely lignified bundle cap layers. They could render the palm optimal bending under high wind loads. Interestingly, the stiffness of the bundle cap layers gradually decreased towards the centre of the stem. This evolution may lead to increased swaying of plants and allow the dissipation of wind energy. These meticulous designs of nature can eventually protect the palm from powerful wind strikes.

旅人蕉 又稱：扇芭蕉

Traveller's Palm, Traveller's Tree | *Ravenala madagascariensis* Sonn.

相片拍攝地點：香港動植物公園
Tree Location: Hong Kong Zoological and Botanical Gardens

名字由來 MEANINGS OF NAME

種加詞 *madagascariensis* 意指其原生於馬達加斯加，此孤島孕育着眾多生命。由於其莖部具大量水分，能為旅客提供應急水源，故得俗名 Traveller's Palm（中文意譯為旅人蕉）。其巨大的葉片水平展開，形成扇狀樹冠。傳聞旅人蕉的葉片總是向南北軸線展開，猶如旅行者的指南針，為旅行者引導正途，不過目前並無可靠證據證明其真確性。另一個與旅行者有關的故事源自於旅人蕉的吸水能力。旅人蕉的葉鞘儲存了大量水分，當旅行者在路途中缺水或口渴時，可直接砍下葉鞘飲用內部的水分。

本地分佈狀態 DISTRIBUTIONS	外來物種 Exotic species
原產地 ORIGIN	馬達加斯加。 Madagascar.
生長習性 GROWING HABIT	常綠，棕櫚狀多年生植物。高度可達 6 米。樹冠對稱，呈扇狀。 Evergreen, palm-like perennial herb. Up to 6 m tall. Crown symmetrical, fan shaped.

1	2	3	4	5	6	7	8	9	10	11	12	花果期 月份

花期：本港全年可見。果期：本港全年可見。
Flowering period: January to December in Hong Kong. Fruiting period: January to December in Hong Kong.

The specific epithet *madagascariensis* indicates its origin from Madagascar, an isolated island nourishing a myriad of precious creatures. The common name "Traveller's Palm" comes from the fact that its stem has a lot of water, which provides emergency water sources for travelers. The huge leaves spread out horizontally, forming a fan-like crown. Rumour has it the leaves always spread along the north-south axis and serve a compass for travellers, but there is missing any testament to the words. Another story is derived from its water capturing ability. Traveller's Palm retains an appreciable amount of water inside the leaf sheaths. It serves as an immediate water source for thirsty travellers who can chop down a leaf sheath and drink the water inside.

應用 APPLICATION

旅人蕉的形態猶如孔雀開屏，姿態華麗，故被廣泛種植在公園和花園中。香港迪士尼樂園就種植了不少旅人蕉迎客，引導來訪者前往樂園的酒店大堂。

Traveller's Palm is predominantly planted in parks and gardens in respect of its showy appearance. Hong Kong Disneyland has planted it as an informative species to direct visitors to the lobby of Hong Kong Disneyland Hotel.

辨認特徵 TRAITS FOR IDENTIFICATION

① 樹幹 TRUNK	② 樹皮 BARK	③ 葉 LEAVES
④ 花 FLOWERS	⑤ 果 FRUITS	

① 旅人蕉的樹幹。

Trunk of *Ravenala madagascariensis* Sonn.

② 主幹棕櫚狀，木質，直立，具明顯環狀葉痕。

Trunk palm-like, woody, erect, with notable leaf-scar rings.

③ 葉片在莖部頂端排成 2 列，大型葉片，長橢圓形，向圓形頂端漸狹，基部近截形，葉緣全緣，兩面顏色一致。葉柄較長，基部葉鞘中空。

Leaves arranged in 2 ranks toward top of the stem. Blade large, oblong, narrower toward the rounded apex, base subtruncate, margin entire, concolorous on both surfaces. Petioles long, basal leaf sheath hollow.

④ 蠍尾狀聚繖花序，腋生，具 6-12 個大苞片，船狀，互生和離生。每個苞片具 5-12 朵小花，呈白色，萼片 3 片，花瓣 3 片。

Scorpioid axillary, bracts 6-12, large, boat-like, arranged alternately and distichously. Each bract holding 5-12 small flowers, white, sepal 3, petals 3.

⑤ 蒴果，木質，具 3 裂。種子眾多，呈黑色，覆有具光澤的藍色假種皮。

Capsules woody, 3-lobed. Seeds numerous, black, covered with glossy blue aril.

生態 ECOLOGY

在馬達加斯加，野生旅人蕉主要依靠領狐猴（又稱黑白領狐猴）授粉。領狐猴是人見人愛的馬達加斯加特有種，擁有熊貓般的毛色和浣熊般的尾巴。牠們十分依賴植物上的蜜源。領狐猴為取得花蜜而打開花苞片時，花粉會黏附在牠們身上，當牠們移動至其他旅人蕉覓食時，便會意外為旅人蕉授粉。

旅人蕉的種子被假種皮包覆。假種皮是一層由珠柄發展而成的附屬物，並非如種皮由珠被發展而成。艷麗的假種皮可引起傳粉者的注意。旅人蕉的假種皮就呈藍色，能成功吸引色盲的領狐猴留意。

在非法伐林和採礦等威脅下，領狐猴的數量大幅減少，現時已被列為極危物種。隨着野生傳粉者的數量減少，野生旅人蕉在汰弱留強的大自然中無可避免地逐漸被淘汰。

In Madagascar, the wild Traveller's Palm is mainly pollinated by *Varecia variegata* (Black-and-white Ruffed Lemur). Black-and-white Ruffed Lemur is an adorable endemic species of Madagascar, with panda-like complexion and a raccoon-like tail. They rely greatly on the nectars of the plant. While they open the bracts for the nectars, the pollen grains attach to their skins and are pollinated during their forage to another Traveller's Palm.

Traveller's Palm's seeds are covered with aril, a layer of appendage that is not derived from integuments as a seed coat, instead, is developed from funiculus. Aril is always showy and can draw the attention of dispersers. In the case of Traveller's Palm, the aril is blue and appealing to Black-and-white Ruffed Lemurs which are protanopia.

Threatening by illegal logging and mining, however, the population of Black-and-white Ruffed Lemur has declined drastically and is currently classified as a Critically Endangered species. Following the reduction of wild pollinators, the wild population of Traveller's Palm could not shun from elimination as well.

生命力 VITALITY

旅人蕉偏愛潮濕、日照充足、排水及通氣良好的土壤。

Traveller's Palm prefers moist, sufficient sunlight and soils with well aeration and drainage.

致謝

Acknowledgements

在此衷心感謝香港特別行政區政府發展局綠化、園境及樹木管理組，對此項目的支持（項目編號：WQ/083/21），並提供 100 種景觀樹木的分佈地點及相關資料，令此項目及本圖鑑能順利完成出版。

We would like to express our sincere thanks to Greening, Landscape and Tree Management Section, Development Bureau, the Government of the Hong Kong Special Administrative Region, for the support of this project (Project No.: WQ/083/21), and for providing the distribution locations and relevant information of these 100 ornamental trees, which made the project and this photographic guide completed and published successfully.

樹藝知識進修

Further Study of Arboriculture

　　如果有興趣進一步了解植物科學和樹藝知識，甚至希望從事園藝及樹藝監督、樹木評估、園境管理，以及園境發展和環境保育等工作，香港高等教育科技學院（THEi）提供園藝樹藝及園境管理（榮譽）理學士課程，為未來的樹藝師、樹木工程監督、園藝或樹木專家、園藝或園景顧問及教育工作者等，培養所需的專業技能和豐富有關知識。

　　If you are interested in learning more about plant science and arboriculture, or even plan to pursue a career in horticultural and arboricultural supervision, tree assessment, landscape management, as well as landscape development and environmental conservation, etc., the Technological and Higher Education Institute of Hong Kong (THEi) offers the programme of Bachelor of Science (Honours) in Horticulture, Arboriculture and Landscape Management. The programme equips the future arborists, tree works supervisors, horticulture and tree specialists, or landscape consultants and educators, etc. with professional skills and knowledge.

香港高等教育科技學院（THEi）

園藝樹藝及園境管理（榮譽）理學士
Bachelor of Science (Honours) in Horticulture, Arboriculture and Landscape Management

課程詳情 Programme Details：

香港 100 種
景觀樹木圖鑑

主編
張浩

副主編
李佳鎮、邱佩晴、黃子衡、洪文君、胡新月

編委
梁彥姿、張綺文

責任編輯
李欣敏

裝幀設計
羅美齡

排版
辛紅梅

出版者
萬里機構出版有限公司
香港北角英皇道 499 號北角工業大廈 20 樓
電話：2564 7511　　傳真：2565 5539
電郵：info@wanlibk.com
網址：http://www.wanlibk.com
　　　http://www.facebook.com/wanlibk

發行者
香港聯合書刊物流有限公司
香港荃灣德士古道 220-248 號荃灣工業中心 16 樓
電話：2150 2100　　傳真：2407 3062
電郵：info@suplogistics.com.hk
網址：http://www.suplogistics.com.hk

承印者
中華商務彩色印刷有限公司
香港新界大埔汀麗路 36 號

出版日期
二〇二三年七月第一次印刷
二〇二四年七月第二次印刷

規格
16 開（240 mm × 170 mm）

鳴謝
香港高等教育科技學院（THEi）
香港園境承造商協會（HKGCA）
香港園藝專業學會（HKIHS）
動力資源有限公司